RENEWABLE HYDROPOWER TECHNOLOGIES

Edited by **Basel I. Ismail**

Renewable Hydropower Technologies
http://dx.doi.org/10.5772/63684
Edited by Basel I. Ismail

Contributors

Ramesh Bhatt, Jacson Hudson Inácio Ferreira Ferreira, José Roberto Camacho, Divas Karimanzira, Edwin Chica, Ainhoa Rubio Clemente, Basel I. Ismail

Published by InTech

Janeza Trdine 9, 51000 Rijeka, Croatia

Publishing Process Manager Ana Simcic
Technical Editor SPi Global
Cover InTech Design team

First published July, 2017
Printed in Croatia
Legal deposit, Croatia: National and University Library in Zagreb

Additional hard copies can be obtained from orders@intechopen.com

Renewable Hydropower Technologies, Edited by Basel I. Ismail
p. cm.
Print ISBN 978-953-51-3381-0
Online ISBN 978-953-51-3382-7

For EU product safety concerns:
IN TECH d.o.o., Prolaz Marije Krucifikse Kozulić 3, 51000 Rijeka, Croatia,
info@intechopen.com or visit our website at intechopen.com.

Meet the editor

Dr. Basel I. Ismail is currently an associate professor and chairman of the Department of Mechanical Engineering, Lakehead University, Thunder Bay, Ontario, Canada. In 2004, Professor Ismail earned his PhD degree in mechanical engineering from the McMaster University, Hamilton, Ontario, Canada. From 2004 to 2005, Dr. Ismail worked as a postdoctoral researcher at Engineering Physics Department, McMaster University. His specialty is in engineering heat transfer, engineering thermodynamics, and energy conversion and storage engineering. Dr. Ismail's research activities are theoretical and applied in nature. Currently, his research areas of interest are primarily focused on green engineering technologies related to alternative and renewable energy systems for power generation, heating, and cooling. This includes applications and innovations in solar energy, wind energy, hydropower, geothermal and ocean energy, thermoelectricity, and other similar areas. Dr. Ismail also has interest in other engineering-based research areas, such as energy efficiency, optimization of thermal and power systems, energy and exergy analyses of energy systems, solar-/wind-powered desalination systems, carbon dioxide membrane gas separation (for industrial emission control and GHG reduction), real-time neutron radiography, and automotive diesel EGR cooling systems. Dr. Ismail was the leading research investigator in a collaborative project (2007–2010) with Goldcorp-Musselwhite Canada Ltd. and Engineering of Lakehead University. This innovative project was state of the art in geothermal energy–related technology applied in Northwestern Ontario, Canada. Dr. Ismail has published many technical reports and articles related to his research areas in reputable international journals and conferences and participated in international forums related to renewable energy technologies. During his research activities, Professor Ismail has supervised and trained many graduate students and senior undergraduate students in mechanical engineering with projects and theses related to innovative renewable and alternative energy engineering and technologies.

Contents

Preface

Hydropower has always been considered as one of the most efficient and magnificent sources of power gneration. For many years, hydropower played an essential role in the development of humanity and has a long and successful track record. It is a conventional renewable energy source for generating electricity in small and large scale production. The power generation from hydroelectric power plants can exceed 10 GW.

This book is the result of contributions from several researchers and experts worldwide. An introductory chapter is initially presented in this book whcih will highlight the principles of hydroelectric power generation and presents a summary of the significance of various aspects related to the following book chapters. Due to its important utilization and future prospects, various interesting topics of research related to hydroelectric power generation are covered in this book. The topics cover the following important aspects: Prospects of hydropower technology; Design of zero head turbine for power generation; Planning of hydropower production under resources and system knowledge uncertainty; Hydropower impacts and implications with application to a case study;

It is hoped that the book will become a useful source of information and basis for extended research for researchers, academics, policy makers, and practitioners in the area of renewable hydropower technologies.

I would like to thank all chapter authors for their efforts and the quality of the chapters presented. Also, I would like to thank Ms. Ana Simcic from InTech-Open Science for her outstanding efforts in managing the publication process of this book.

This book contains five chapters. First chapter is an introduction on aspects of renewable hydroelectric power generation. Chapter two discusses interesting prospects of small hydropower technology. Small Hydropower technology is one of the most common technologies used for electricity generation for rural population in both developed and developing countries. Inclusion of the remains of this resource in the energy mixes could lead to sustainable development. Small hydroelectric power plants contribute to meeting the needs of regions where there is no a major technological development and they are able to improve the population's quality of life with the creation of jobs, increase the local economy and enhancement of the region.

Chapter three covers useful and practical aspects on the design of zero head turbines for power generation in hydropower systems. The blades represent a vital component of any hydrokinetic turbine due to their complexity, cost and significant effect on the operating efficiency. During its lifetime, a hydrokinetic turbine blade is subject to different types of loads, such as hydrodynamic, inertial, and gravitational forces. The hydrodynamic design provided the blade external shape; i.e., the chord and twist angle distributions along the blade, which resulted in optimal performance of the hydrokinetic turbine over its lifetime. The blade element method (BEM) is used in this chapter for the hydrodynamic design of the rotor of a horizontal axis hydrokinetic turbine of 1 kW. Failure analysis of the blades in the rotor is presented.

Chapter four primarily discusses planning hydropower production of small reservoirs under resources and system knowledge uncertainty. Available energy from water varies widely from season to season, depending on precipitation and streamflows, especially in small catchments. In addition, the reservoir operation problem is associated with the inability of operators to formulate crisp boundary conditions, due to uncertainty in knowledge. In this chapter, an approach for planning the operation of small multipurpose reservoir systems for hydropower generation and flood control under consideration of the stochastic nature of inflows and initial storage levels and allow formulation of constraints with some range of uncertainty will be presented. The approach is based on joint chance constrained and fuzzy programming, which addresses the problem of including risk directly in the optimization. Besides the optimal reservoir release strategy, this approach also determines the optimal reliabilities of satisfying hydropower demand and flood control storage requirements. Therefore, this tool has some advantages in planning the operations of reservoirs in extreme hydrological events such as floods and droughts. In this chapter, the system is applied to the Wuyang Small Hydropower plants cascade in the People's Republic of China.

Finally chapter five, the impacts and implications of hydropower in Nepal is presented and discussed. Although Nepal is one of the poorest countries in terms of economy, it has endowed high potential of renewable water resources having 6000 rivers with annual runoff of 170 billion m^3 flowing from mountains to hills and plains covering of 395,000 ha (48%) area in 45,000 km in length are feasible for generation of 45,610 MW hydroelectricity out of 83, 290 MW theoretical potential of the country. Despite the high potential of hydropower potential, Nepal's low economy and slow GDP growth rate in combination with environmental and socioeconomic constraints, effective implementation of existing policy and political stability may support to reach the sustainable development goals of the county.

Dr. Basel I. Ismail, P.Eng.
Associate Professor and Chairman
Department of Mechanical Engineering
Faculty of Engineering
Lakehead University
Thunder Bay, Ontario, Canada

Chapter 1

Introductory Chapter - Aspects of Renewable Hydroelectric Power Generation

Basel I. Ismail

Additional information is available at the end of the chapter

http://dx.doi.org/10.5772/67563

1. Introduction

For many years, hydroelectric power has been considered as one of the most efficient and magnificent sources of power generation. It played an essential role in the development of humanity. It is a conventional renewable energy source for generating electricity in small- and large-scale production. The power generation from hydroelectric power plants can exceed 10 GW. Hydroelectric power is also the most desirable and has a long and successful track record. Hydropower is the power derived from the energy of falling water (as in waterfalls) or fast running water (as in rivers or long water streams). A schematic showing a cross-section of a typical hydroelectric power facility is shown in **Figure 1**. In this figure, water (the working fluid) is collected in a reservoir just behind a dam. The water accumulation behind the dam is normally dependent on several factors such as the intensity, distribution, and duration of rainfalls; the field moisture capacity of the basin or reservoir soil; and the direct evaporation, transpiration, and ground infiltration. The working fluid is then transported through large pipes (called penstock) to the inlet of one or more hydroturbines at the base of the dam. The hydroenergy available from conversion to mechanical shaft work is the potential energy of the water in the main reservoir. The water makes the hydroturbine rotate by action of the available hydroenergy contained in the water. Nowadays, hydroturbines are made more compact and usually operate at high rates of rotation and with high mechanical efficiency. The mechanical shaft work is then converted into electricity using the electric generator coupled with the hydroturbine shaft in the powerhouse. The power is then regulated using electric transformers before being transmitted through powerlines to the main electric grid.

This book provides a unique collection of recent research topics related to hydropower technology. The topics cover the following important aspects: diagnosability of soft fault in hydropower systems, hydropower impacts and implications with application to a case study, prospects of

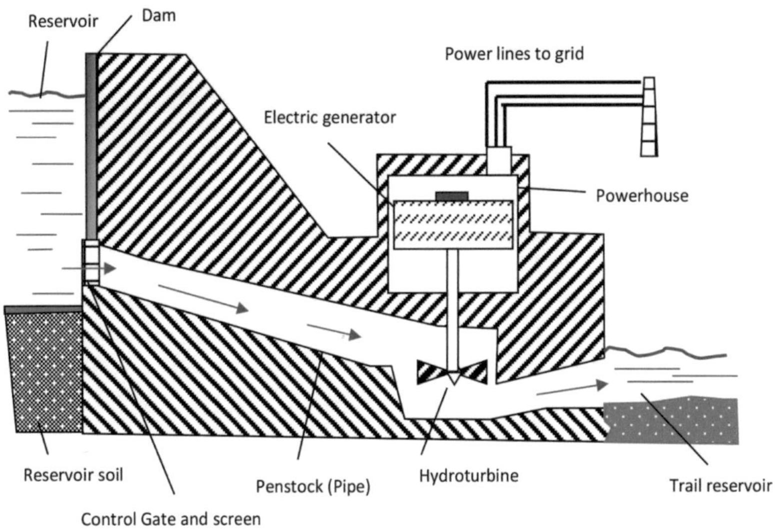

Figure 1. A schematic showing a cross-section of a typical hydroelectric power facility.

hydropower technology, planning of hydropower production under resources and system knowledge uncertainty, and design of zero-head hydroturbines for power generation. Before these practical topics are presented in the following subsequent chapters, it would be useful to provide some introductory energy conversion fundamentals and present brief aspects of concepts of hydroelectric power generation. This is presented in this introductory chapter.

2. Hydropower principles

2.1. The available hydropower

There are two significant operating parameters for hydroelectric power generation potential: (1) the amount of water volumetric flow rate ($\dot{\forall}_w$) and (2) the elevation head that water can be made to fall (h). It should be noted that the elevation head may be attributed to the naturally existing site topography or it may be made artificially by constructing a dam.

The available steady-state ideal hydropower can be estimated using:

$$P_{H,ava,i} = \gamma_w \dot{\forall}_w h \tag{1}$$

where $P_{H,ava,i}$ is the available ideal hydropower (W or kW), γ_W is the specific weight for water ($\cong$9180 N/m^3), $\dot{\forall}_w$ is the available flow of a watercourse (m^3/s), and h is the water elevation head measured between the main and trail reservoirs (m) of the hydropower site.

The available steady-state actual hydropower can be estimated using

$$P_{H,ava,a} = P_{H,ava,i} - P_{HL} \tag{2}$$

where P_{HL} is the hydraulic frictional losses in the piping system (W or kW). These losses can be estimated using

$$P_{HL} = \rho \dot{\forall}_w g \sum H_L \tag{3}$$

where $\sum H_L$ represents the total head losses in the piping system (i.e., penstock) in m. The parameters ρ and g are the density of water and gravitational constant (9.81 m/s^2), respectively. In general, the total head losses are given by:

$$\sum H_L = \sum_{Major\ losses} H_L + \sum_{Minor\ losses} H_L \tag{4}$$

and

$$\sum_{Major\ losses} H_L = \sum_{m=1}^{M} f_m \frac{L_m}{D_m}\left(\frac{V_m^2}{2g}\right) \tag{5}$$

In Eq. (5), f_m is the friction coefficient, and L_m and D_m are the length and internal diameter, respectively, for a given pipe segment m of the long pipe that could have different pipe segments. The flow velocity V can be calculated using

$$V = \frac{\dot{\forall}_w}{A_c} \tag{6}$$

The parameter A_c is the cross-sectional area of the pipe. Substituting Eq. (6) in Eq. (5) and considering a circular pipe yields:

$$\sum_{Major\ losses} H_L = \sum_{m=1}^{M}\left(8f_m \frac{L_m \dot{\forall}_w^2}{g\,\pi^2 D_m^5}\right) \tag{7}$$

The friction coefficient f_m can be estimated depending on the type of flow regime in the pipe, which is characterized by Reynolds number Re. For laminar flow, it is determined using:

$$(f_m)_{Lam} = \frac{64}{Re} \tag{8}$$

and for turbulent flow, it can be estimated using:

$$(f_m)_{Turb} = \frac{0.3086}{\left\{log\left[\left(\frac{\varepsilon}{3.7D_m}\right)^{1.11} + \frac{6.9}{Re}\right]\right\}^2} \tag{9}$$

where ε is the roughness of the inner pipe surface and Reynolds number is defined as:

$$Re = \frac{\rho\, V\, D}{\mu} \tag{10}$$

The minor losses are due to any fittings that might be part of the piping system. It might be given by:

$$\sum_{Minor\ losses} H_L = \sum_{n=1}^{N} K_n \left(\frac{V_n^2}{2g}\right) \tag{11}$$

where K_n is the fitting friction coefficient for a component n in the piping system.

2.2. The hydroelectric power output

The output (produced or generated) of the hydropower depends on the hydroturbine conversion efficiency, the electric generator efficiency, and the electric transformer efficiency. This is given by:

$$P_{H,\,out} = \eta_{HT}\eta_{EG}\eta_{ET}P_{H,\,ava,\,a} \tag{12}$$

where $P_{H,out}$ is the produced (output) hydropower (W or kW), η_{HT} is the hydroturbine energy conversion efficiency, η_{EG} is the electric generator efficiency, and η_{ET} is the electric transformer efficiency. The three types of efficiencies in Eq. (2) can be combined to represent the overall hydropower plant efficiency, η_{HPP}, so that Eq. (2) becomes

$$P_{H,\,out} = \eta_{HPP}P_{H,\,ava,\,a} \tag{13}$$

The total power losses from the plant can be determined using

$$P_{TL} = P_{H,\,ava,\,i} - P_{H,\,out} \tag{14}$$

or using

$$P_{TL} = (1 - \eta_{HPP})P_{H,\,ava,\,a} + P_{HL} \tag{15}$$

Illustrative example A

As an illustrative numerical example, consider a hydroelectric power facility that has a rated electrical capacity of 210 MW. The operating elevation head in this facility is 60 m, and the water flow rate into the hydroturbine is estimated at 510 m^3/s. Estimate (1) the overall efficiency of the hydropower facility and (2) the total power losses from this facility. It can be assumed that the hydraulic frictional losses are negligible compared to other power losses.

Analysis:

The first part of this example can be determined using Eq. (13). Here, $P_{H,out}$=210 MW and $P_{H,ava,a}$ can be estimated using Eqs. (1) and (2):

http://dx.doi.org/10.5772/67563

$$P_{H,ava,a} = 9810 \times 510 \times 60 \times 10^{-6} - 0 \cong 300 \text{ MW} \tag{16}$$

Using Eq. (13):

$\eta_{HPP} = \frac{210}{300} = 0.7$, that is, the hydropower plant is 70% efficient.

The steady-state total power losses from this facility can be estimated using either Eq. (14) or Eq. (15). Using Eq. (14), gives:

$$P_{HL} = 300 - 210 = 90 \text{ MW} \tag{17}$$

Illustrative example B

Consider a microhydropower facility with an available elevation head of 23 m and a pipe of a total length of 350 m and diameter 30 cm. The available discharge of watercourse from the hydro resource is 0.25 m^3/s. Properties of water (taken at standard temperature of 20°C) are ρ=1000 kg/m^3 and μ=0.00114 Pa. s. The hydroturbine efficiency is 85% and the electrical generator and transformer have 100%. The hydropower that could be delivered from this facility is estimated.

Analysis:

Using Eq. (1), $P_{H,ava,i}$ is calculated to be ~56.4 kW, and using Eq. (9), f is estimated to be 0.01411. Applying Eq. (3), P_{HL} = 25.7 kW so that Eq. (2) gives $P_{H,ava,a}$=30.7 kW. Finally, substituting back in Eq. (12) yields $P_{H,out}$=26.1 kW. This means that the microhydroelectric facility is capable of supplying ~26 kW under the given steady-state operating conditions.

3. Some important aspects of hydropower

There are many practical aspects related to renewable hydropower technologies, for example, aspects for consideration in supplying electric power for rural populations in developing countries using small hydropower systems, hydroturbine design aspects, planning hydropower production of small reservoirs under resources and system knowledge uncertainty, impacts and implications of hydropower in economically poor countries, and fault diagnosis and control of operating parameters for optimizing production of hydropower systems. Brief description of these aspects is presented here and more details are covered in the subsequent chapters of this book.

Small hydropower technology is one of the commonly used technologies for electricity generation supplied in rural population in both developed and developing countries. Small hydroelectric power plants contribute to meeting the needs of regions where there is no major technological development, and they are able to improve the population's quality of life with the creation of jobs, increase of the local economy, and enhancement of the region.

The design and innovation of hydroturbines in hydropower systems constitute a major part in the success of this technology. The blades represent a vital component of any hydrokinetic turbine due to their complexity, cost, and significant effect on the operating efficiency. During its lifetime, a hydrokinetic turbine blade is subjected to different types of loads such as hydrodynamic, inertial, and gravitational forces. The hydrodynamic design provided the blade external shape, that is, the chord and twist angle distributions along the blade, which resulted in optimal performance of the hydrokinetic turbine over its lifetime. The blade element method (BEM) could be used for the hydrodynamic design of the rotor of a horizontal axis hydrokinetic turbine of mini power production.

Planning hydropower production of small reservoirs under resources and system knowledge uncertainty is another important practical aspect that should also be considered. Available energy from water varies widely from season to season, depending on precipitation and streamflows, especially in small catchments. In addition, the reservoir operation problem is associated with the inability of operators to formulate crisp boundary conditions due to uncertainty in knowledge. In this chapter, an approach for planning the operation of small multipurpose reservoir systems for hydropower generation and flood control under consideration of the stochastic nature of inflows and initial storage levels that allow formulation of constraints with some range of uncertainty will be presented. The approach is based on a joint chance that is constrained and fuzzy programming, which addresses the problem of including risk directly in the optimization. Besides the optimal reservoir release strategy, this approach also determines the optimal reliabilities of satisfying hydropower demand and flood control storage requirements. Therefore, this tool has some advantages in planning the operations of reservoirs in extreme hydrological events such as floods and droughts. A case study could be considered in using this practical approach.

The impacts and implications of hydropower in economically poor countries are other insightful aspects. Despite the high potential of hydropower, a country's low economy and slow GDP growth rate in combination with environmental and socio-economic constraints, effective implementation of existing policy, and political stability may be supportive to reach the sustainable development goals of this country.

The diagnosability of soft fault in hydropower systems. In a hydropower system, the operating hydroturbine speed is one of the critical variables that requires monitoring and optimization for efficient control of the frequency and output voltage from the electrical generator coupled with the turbine. Diagnoses and control aspects in a hydropower system.

Author details

Basel I. Ismail

Address all correspondence to: baseliai@gmail.com

Department of Mechanical Engineering, Lakehead University, Canada

Chapter 2

Prospects of Small Hydropower Technology

Jacson Hudson Inácio Ferreira and

José Roberto Camacho

Additional information is available at the end of the chapter

http://dx.doi.org/10.5772/66532

Abstract

Small hydropower (SHP) belongs to renewable energy technology group and is a form of attractive power generation environmental perspective because of its potential to be found in small rivers and streams. Many countries use the technology of small hydro as a renewable energy source in order to minimize existing environmental effects in the production of electricity and have the maximum use of water, a renewable resource. This technology has shown prominence on the world stage with seemingly insignificant environmental effects on rivers, water channels, and dams.

Keywords: hydropower, small hydropower, renewable sources

1. Introduction

The renewable energy sources can be auxiliaries in reducing the environmental impacts that energy production using fossil fuels or other nonrenewable resources causes in nature, with its form of clean and sustainable production. Some known forms of alternative generation are hydro, wind, solar, geothermal, and biomass, which are already present in the energy matrix of various countries, and they have highlighted the way governments have proposed to extract energy.

According to the United Nations Industrial Development Organization (UNIDO), the technology applied in small hydropower (SHP) of renewable form allows the development of rural areas and the access to electricity by a portion of the population living in these regions and they contribute to sustainable development and social inclusion. These factors are positive in the evaluation of governments and their public policies [1].

Hydroelectric generation has a high cost of deployment and maintenance, and short-term disadvantage is observed. In the long run, this alternative source becomes attractive for both clean and sustainable generation and advantage of being exploited too close to large consumer centers, reducing costs of distribution for example.

2. Small hydropower

2.1. Definition and classification

There is no an internationally agreed definition for a small hydropower plants, and its classification is based only on a country's level of hydropower development. **Table 1** shows the definition and classification in some countries with prominent in the generation of electricity by small hydropower in the world.

Country/organization	Micro (kW)	Mini (kW)	Small (kW)
Brazil	<100	101–1000	1001–30,000
China	≤100	≤2000	≤50,000
Philippines	-	51–500	<15,000
Sweden	-	-	101–15,000
USA	<500	501–2000	<15,000
India	<100	<2000	-
Japan	-	-	<10,000
Nigeria	≤500	501–2000	-
France	<500	501–2000	<50,000
New Zealand	-	<10,000	<50,000
United Kingdom	<1000	-	-
Canada	-	<1000	1001–1,500
Russia	-	-	<30,000
Norway	<100	101–1000	1000–10,000
Germany	<500	501–2000	<12,000
Turkey	<100	101–2000	<10,000

Source: Ref. [2].

Table 1. SHP definition and classification in some selected countries.

Hydropower plants are of three types [3]:

- Impoundment: this hydropower system is applied to large generation where water is accumulated in its reservoir, using the dam system.

- Diversion: for the generation of electricity with diversion is necessary to build a canal or penstock for a part of the river can go to the generating group. The dam system cannot be required for the diversion system.
- Run-of-river: this system utilizes the natural flow of water in the river and in some situations do not need impoundment.

The choice of small hydropower plant technology is based on the use of the system of run-of-river so that there is little or no dam on the site that owns the hydroelectric project, using the kinetic energy of moving water to move turbines. The system run-of-river decreases the negative effects that the large hydropower plant causes in the plant installation region as the flooding of arable land and disturbances in the temperature and composition of the river [4].

2.2. Characteristic and components

Figure 1 illustrates a typical run-of-river small hydro scheme.

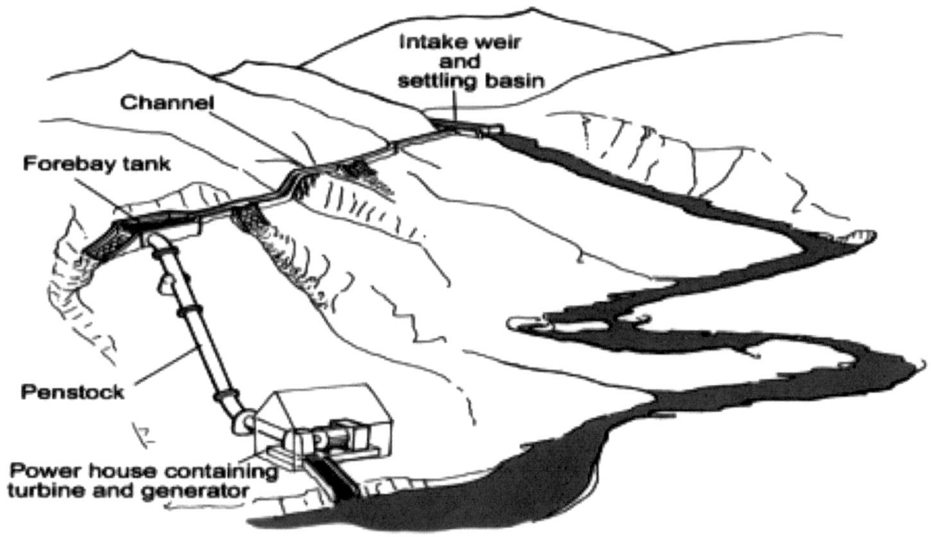

Figure 1. Typical small hydro site layout. Source: Ref. [5].

The fundamental elements are the weir, the settling tank (the forebay), the penstock, and a small canal or "leat." Water is diverted from the course (main river) through an intake at the weir. The weir is a man-made barrier cross the river, which regulates the water flow through the intake. Before entering the turbine, the particulate matter is removed by passing water through a settling tank. Water is sufficiently slowed down in the settling tank for the particulate matter to settle out. A protective rack of metal bars (trash rack) is typically found near the forebay to protect the turbines from damage by larger materials such as stones, timber, leaves, and man-made litter that may be found in the stream [3].

To understand the factors that affect the benefits of an SHP first is necessary to understand the role that the major components have on a hydroelectric project. Stand out for small hydroelectric power stations the following components [6].

- Dam is the structure of a plant responsible to elevate and keep the level upstream of the engine room, creating artificially a local unevenness.
- Spillway designed in order to drain the greater design flow for the maintenance of the reservoir-required water level, avoiding the risk of the water reaches the dam crest. It is the dam safety structure.
- Generation circuit consists of channels water intakes, pipes or low-pressure adduction tunnels, any surge shafts or load chambers, high-pressure ducts or forced tunnels, external or underground powerhouse, tunnels, and leakage channels. The generation circuit is intended to adduce water for the transformation of mechanical energy into electrical energy.

For the generation circuit we have:

- Water intake: structure to capture the water to the penstock or channel/adduction tunnel.
- Channel and adduction tunnel: structures responsible for adduct the water to the forced conduct in shunt arrangements.
- Equilibrium chimney: aims to stabilize pressure changes resulting from partial or total change of water discharge in starting conditions, load variations, or load shedding of a generating unit.
- Load chamber: is the structure that makes the transition between the channel and the water intake of the penstock. It is dimensioned in order to meet critical starting conditions and sudden stop of the generating unit.
- Penstock: is the structure that connects the water intake to the powerhouse working under pressure. The penstocks can be external or tunnels.
- Powerhouse: structure that houses the electrical and mechanical equipment. The typical powerhouse arrangement is as in any project of this nature, conditioned by the type of turbine and generator.
- Tunnel or tailrace: located downstream of the suction tube between the powerhouse and the river, is the channel through which the turbinated water is discharged and returned to the river.

2.3. Project steps

The use of a hydroelectric potential is an activity subject to institutional, environmental, and commercial regulations. Throughout the project implementation process, multidisciplinary activities are mixed, constituting the legal framework of the entire project. **Flowchart 1** shows the activities that are typical for the development and study of an SHP, depicting the interdisciplinary of studies.

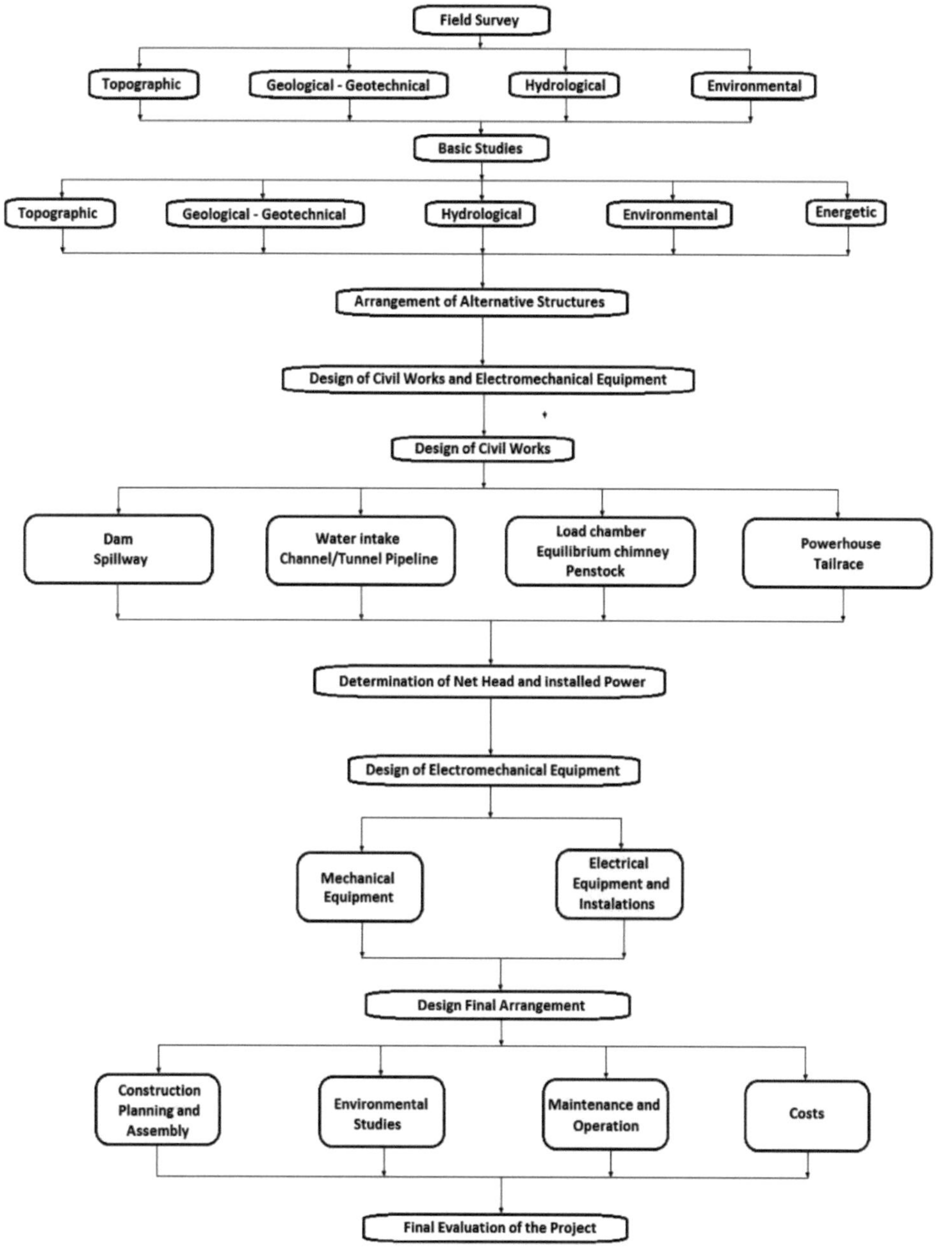

Flowchart 1. Studies and SHP projects. Source: Ref. [7].

The implementation of a project, which aims to use a hydroelectric project for power generation, has a step cycle including phases that estimate, plan, and execute the project. Based on reference [7], these phases are:

- Estimation of hydropower potential: this step is carried to the preliminary analysis of the characteristics of the river basin, especially with regard to topographic, hydrological, geological, and environmental aspects, in order to verify their vocation to generate electricity. This analysis exclusively guided in the available data, is done in the office and allows the first assessment of the potential and cost estimate of the utilization of the watershed and the priority setting to the next step.

- Hydroelectric inventory: it is characterized by the design and analysis of various falling division alternatives for the river basin, formed by a set of projects, which are compared to each other, in order to select the one that presents the best balance between deployment costs, energy benefits, and environmental impacts. This analysis is done based on secondary data, supplemented with field information, and guided by basic studies cartographic, hydro, energy, geological and geotechnical, environmental, and multiple water uses. This analysis will result in a set of exploitations, its main features are indexes, cost/benefit, and environmental indices. It is part of inventory studies submit alternative utilizations selected a study of integrated environmental assessment in order to support the licensing process. These exploitations then become included in the list of inventoried utilizations of the country, capable of composing the expansion plans described above.
- Viability: here the studies are more detailed to the analysis of technical, energy, economic, and environmental viability leading to the definition of the optimum use that will be to power auction. The studies include field investigations on site and include the design of the use of the reservoir and its area of influence and the works of local and regional infrastructure necessary for its implementation. Incorporate analysis of the multiple uses of water and environmental interference. Based on these studies, we are prepared the environmental impact assessment (EIA) and environmental impact report (EIR) of an enterprise specific, with a view to obtaining the preliminary license (PL) with environmental agencies.
- Basic design: the design in the feasibility studies is detailed in order to define more precisely the technical characteristics of the project, the technical specifications of civil and electromechanical equipment, as well as social and environmental programs. The basic environmental project should be prepared for detailing the recommendations contained in the EIA in order to obtain the installation license (IL) for the contracting of works.
- Executive project: includes the preparation of drawings detailing the civil works and electromechanical equipment necessary for executing the works and installation of equipment. At this stage all appropriate steps are taken for the implementation of the reservoir, including the implementation of environmental programs, to prevent, mitigate, or compensate for environmental damage and should be required to operating license (OL).

2.4. Costs of the project

It is important to analyze the existing conditions at the installation site of a hydroelectric plant in order to minimize the installation costs and maximize power generation. Installation costs vary according to the region of installation, infrastructure, and generation capacity. The equipment is also part of the factors that increase the cost of a plant. The small plants also have high installation costs even with a smaller size [3].

According to Forouzbakhsh et al. [8] and Hosseini et al. [9], during the project study phases of an SHP, it is necessary to divide all the construction costs, operation, and maintenance in two categories: investments and annual costs. Investment costs include the electrical and mechanical equipment, transmission towers, civil structures, and other costs classified as indirect. Already the annual costs are the necessary with maintenance, operation, prevention, and replacement of components and equipment [8, 9].

2.4.1. *Investment costs*

Direct costs include civil costs, electro-mechanical equipment costs, and power transmission line costs as listed below [8, 9]:

- Civil costs are calculated for the structural aspects of a design and construction of the plant, this includes the dam, forebay, tailrace channel, and penstock, among other aspects that are designed in the feasibility stage of a project.
- Generators, turbines, control systems, substations, protective equipment and actuation, and other electrical equipment belong to the costs of electromechanical equipment during the planning phase of an SHP. The costs associated with electromechanical equipment of an SHP can change according to the potential of the plant.

The cost of electromechanical equipment can also be determined using the power, P and the net head, H of the small hydropower plant from [10]:

$$\text{Cost} = aP^{b-1}H^{c}(€/\text{kW}) \tag{1}$$

where a, b, and c are coefficients that depend on the geographical, space, or time field where they are being used.

- The transmission line costs include the lines from the generation stage until the arrival of energy in the substation. These costs depend on the location, infrastructure, highways, existing systems, and the generation capacity of SHP. However, the value is high as the size of the transmission line increases.

According to Hosseini et al. [9], the indirect costs include engineering and design (E&D), supervision and administration (S&A), and inflation costs during the construction period [9].

- E&D costs: the parameters such as location and size of the project can change the E&D costs. These costs are analyzed as a percentage of construction costs, along with the equipment and civil works. These factors are different from one region to another. Studies show that the plants with small potential, the cost can range from 5% and it varies in the plant with great potential to 8%.
- S&A costs: the acquisition of land, the cost of management activities, supervision, and inspection belong to S & A costs. This cost is similar to the cost of E & D and is also analyzed as a percentage of construction costs. The values can range from 4 to 7% depending on the installation site.
- The inflation rate during all the project phases must be taken into consideration. Deployment costs must be adjusted to the inflation rate of the period and of the next few years, determined by the average inflation rate of the previous years.

2.4.2. Annual costs

To obtain the net benefit of a project, annual costs, in addition to investment costs should be calculated. Annual costs include depreciation of equipment, operating and maintenance (O&M), and replacement and renovation costs [9].

- Depreciation of equipment: the service life, wear, and factors that may change the operation of the equipment need to be analyzed during the economic planning of the project.
- O&M costs: the amounts spent on professionals in an SHP project, such as salary, insurance, taxes, and consumables, are attached to the annual costs. These expenses are corrected with the annual inflation local. It is used on a 5% inflation rate for the correction of the costs of the professionals. These costs represent 2% of total annual investment costs.
- Replacement and renovation costs: some items of the main electromechanical components and of great importance as generator windings and turbine runners will need to be replaced or exchanged sometimes. It is believed that the costs required for maintenance and repairs of equipment will have the same value as the total elapsed equipment after 25 years. Therefore, equipment wear costs should be determined for each component or equipment needed in the power generation process in an SHP.

2.5. Principles

The basic hydropower principle is based on the conversion of a large part of the gross head, H(m) into mechanical and electrical energy. Hydraulic turbines harness the water pressure to convert potential energy into mechanical power, which drives an electric generator and other machines. The energy that has water is directly proportional to the pressure and flow. **Figure 2** shows several components of an SHP.

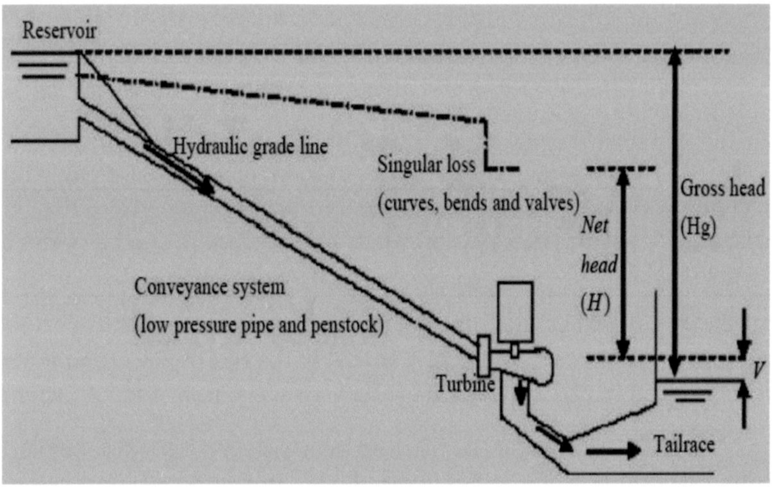

Figure 2. Components of a small hydropower. Source: Ref. [11].

Generally, the hydraulic power P_0 (kW) and the corresponding energy E_o (kWh) over an interval of time Δt (h) are

$$P_0 = \rho g Q H \tag{2}$$

$$E_0 = \rho g Q H \Delta t \tag{3}$$

where ρ and g are the density of water (kg m^{-3}) and Q acceleration due to gravity (ms^{-2}), respectively. The final power, P delivered to the network is smaller than P_0. The power output of any hydropower plants is given by

$$P = \eta P_0 \tag{4}$$

where η is the hydraulic efficiency of the turbo-generator.

First, to set the type of hydraulic turbine of a project, it is observed the head and the volume of water in the river or the local plant installation. However, for a complete and objective analysis for choosing the turbine must consider the efficiency and cost of each existing type of turbine. There are two types of the turbines for the SHP [6]:

- Impulse turbine—pelton, cross flow, and turgo.
- Reaction turbine—propeller, Francis, and Kaplan.

3. Turbine selection

Hydro turbines can be categorized into two groups: impulse turbines and reaction turbines. The difference relates to the way that energy is produced from the inflows [3].

3.1. Impulse turbine

Impulse turbine uses the kinetic energy of water to drive the runner and discharges to atmospheric pressure. The runner of impulse turbines operates in air and is moved by jets of water. According Okot [3], "the water that falls into the tail water after striking the buckets has little energy remaining, thus the turbine has light casing that serves the purpose of preventing the surroundings against water splashing." This type of turbine has great applicability in systems with a large falling water and low flow. Three types of impulse turbines are common in power plants: the pelton, the cross flow, and turgo [3].

3.1.1. Pelton

The pelton turbine (**Figure 3a**) has a high operating head. Because the operating head is so high, the flow rate tends to be low, amounting to as little as 0.2 cfs. The turbine requires the flow through the inlet to be highly pressurized, making the proper penstock design crucial. The pelton utilizes a nozzle located in the spear jet, which is used to focus the flow into the buckets on the runner. The spear jet and buckets are designed to create minimal loss; this leads to a potential efficiency of 90%, even in small hydro applications. A pelton turbine can have up to six spear jets (shown in **Figure 3b**), which effectively increase the flow rate to the turbine resulting in a greater power production and efficiency [12].

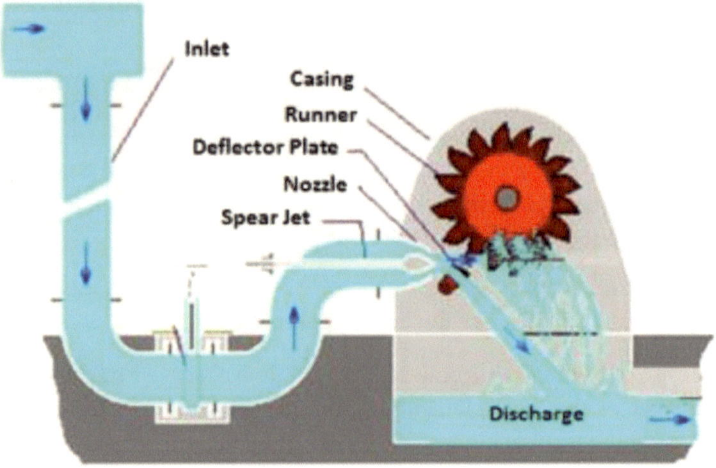

(a)

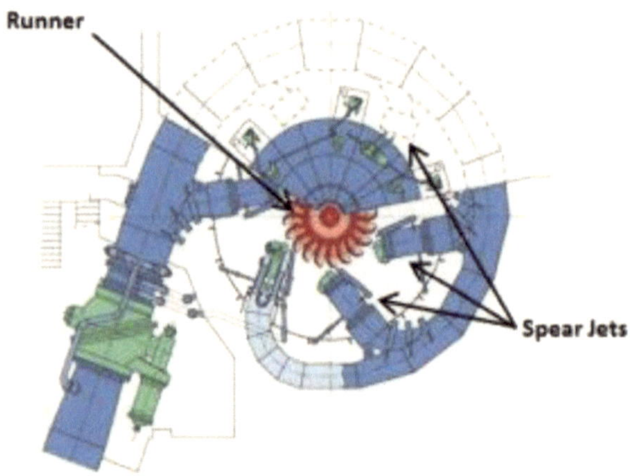

(b)

Figure 3. (a) Pelton turbine schematic. (b) Cross section of pelton turbine. Source: Ref. [12].

3.1.2. Cross flow

The cross-flow turbine (**Figure 4**) is named for the way the water flows across the runner. This is because several cross flow in its construction have at its entrance two or more inlet guide vanes. This impulse turbine class displays for a variety of flow rates at high efficiency. By altering the operation of the inlet guide vanes to better suit flow conditions, flow can be directed at just a portion of the runner during low inflow, or the entire runner when higher flows dictate. As evident from the efficiency curve, the cross flow is able to maintain a consistent efficiency [12].

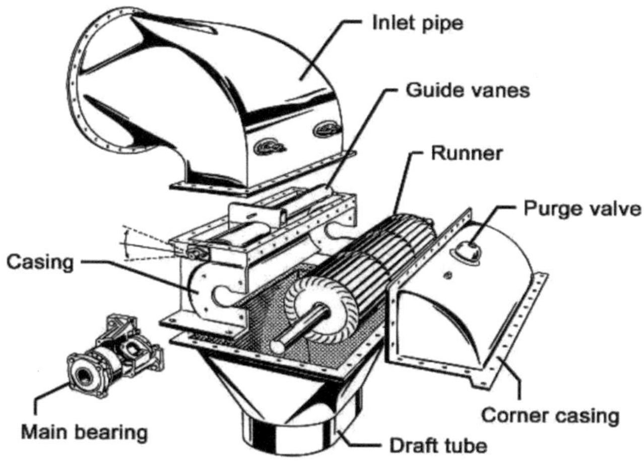

Figure 4. Cross flow schematic. Source: Ref. [12].

3.1.3. Turgo

This turbine is similar to the pelton, but with different shape of the buckets and the jet strikes the plane of the runner at an angle. The turgo turbine has some differences compared to the pelton turbine that in some projects your application can be favorable. This turbine has a high overall efficiency and low maintenance due to its running speed being higher, allowing a direct coupling most likely between the turbine and generator. The turgo turbine can have a smaller diameter compared with the pelton because the flow rate passing through it is not limited as input discharged jet, increasing energy production [3].

3.2. Reaction turbine

The turbines of reaction generate energy by mutual activity of pressure and flow of water. They operate when the rotor is involved in a casing pressure and is completely under water. According to Okot [3], "the runner blades are profiled so that pressure differences across them impose lift forces, akin to those on aircraft wings, which cause the runner to rotate." Unlike impulse turbines, reaction turbines are applicable in places with low height drop and higher flow of water. Examples are the turbines: propeller, Francis, and Kinetic [3].

3.2.1. Propeller

Okot [3] claims that a propeller turbine has a generally axial flow passageway of three to six blades, depending on the designed water head. For efficiency, the water needs to be given a swirl before entering the turbine hall. Sites with low flow of water are suitable for propeller turbines. Bulb, Kaplan, and Straflo are examples of propeller turbines.

Also according to Okot [3], "for adding inlet swirl include fixed guide vanes mounted upstream of the runner and a snail shell housing for the runner, in which the water enters tangentially and is forced to spiral in to the runner.. In the case of Kaplan turbine runner blades are set." Adjusting the turbine blades and guide vanes can significantly improve efficiency in a wide range of flows; however, it is expensive and so can only be economical in larger systems. Where there is potential for small plants and flow and waterfall are somewhat constant, propeller turbines unregulated are commonly used. **Figure 5** shows a typical propeller turbine.

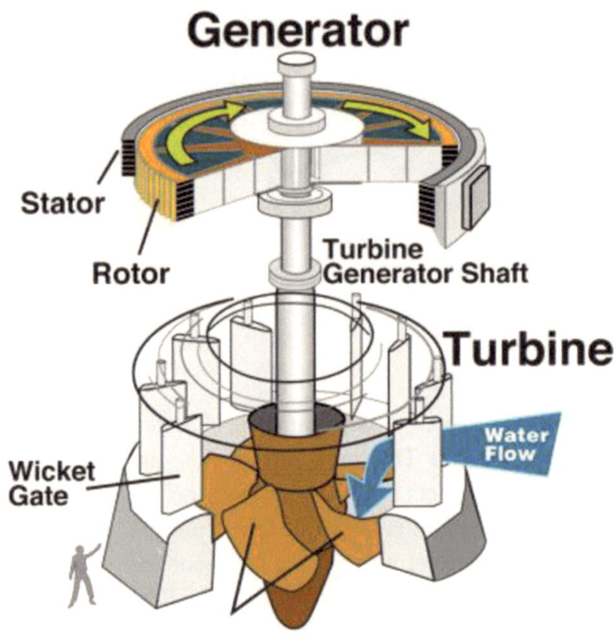

Figure 5. Propeller schematic. Source: Ref. [3].

3.2.2. Francis turbine

One of the more classical designs of hydraulic turbines, the Francis turbine has an efficiency curve, which can function in different situations of height and flow, and its constructive aspects have adjustable guide vanes and fixed runner blades [12].

For this turbine (**Figure 6**), Okot [3] claims that the turbine "generally has radial or mixed radial/axial flow runner which is most commonly mounted in a spiral casing with internal adjustable guide vanes. Water flows radially inward into the runner and emerges axially, causing it to spin. In addition to the runner, the other major components include the wicket gates and draft tube."

Francis turbines can have an efficiency of 90% when the project height is average but can be inefficient when the flow measured at the site is very different from the design flow. This turbine can be set to an open trough or be attached to a penstock [3].

3.2.3. Kinetic

Kinetic turbines harness the kinetic energy of water to produce energy, i.e., it takes advantage of the natural flow of water. Thus, the kinetic systems do not use deviations or artificial channels; they use the natural course of the river. However, they can be applied in such conduits [3]. **Figure 7** shows a type of kinetic turbine.

(a)

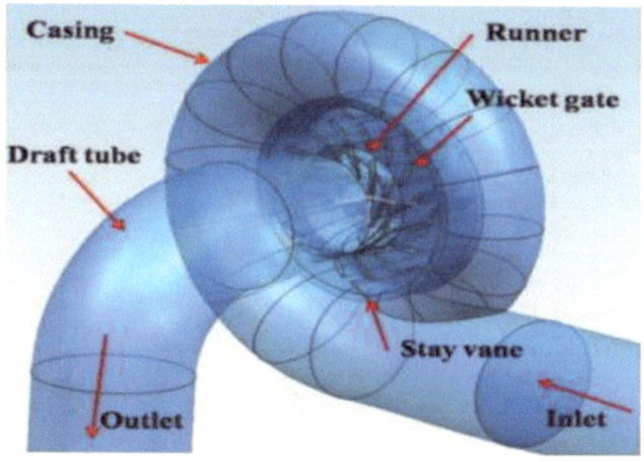

(b)

Figure 6. (a) Francis runner. (b) Francis schematic. Source: Ref. [12].

Figure 7. Hydrokinetic model. Source: Ref. [13].

3.3. Selection

In small hydroelectric plants, determining the type of turbine that will be able to operate, as the design data, it can be made in standard sizes turbine, although there is a difference in terms of efficiency. The chart (**Figure 8**) below shows seven major types of turbines and their recommended range of head and flow [12].

The graph of **Figure 8** can be used in the stage of preliminary studies to choose which type of turbine according with the project height and the water flow in the plant installation site and analyze their hydroelectric potential. According to the chart, if the local is identified at a height of 100 feet and a flow of 100 cfs, the turbines of Kaplan, Francis, and cross flow are presented as options, each with its advantages and disadvantages.

The hydraulic turbine manufacturers offer an efficiency curve. This curve shows the relationship between flow and fall of water and how efficiently they are analyzed according to the results of these two variables. It is possible to analyze each type of turbine and its comportment in the several situations of the project. Generally, a flatter efficiency curve represents a turbine that can operate under broad ranges of head and flow. Curves that are steeper and narrower are indicative of a turbine designed for more focused ranges of operation. **Figure 9** shows the turbine efficiency chart.

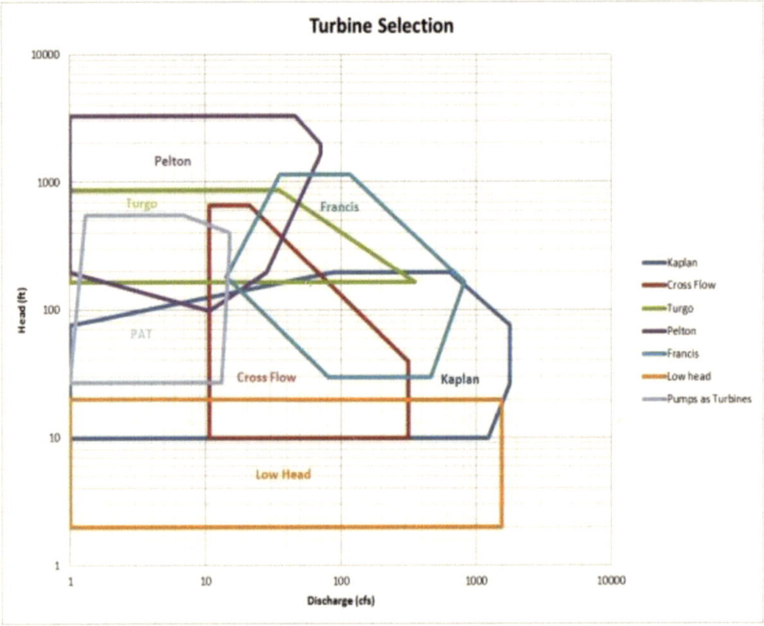

Figure 8. Turbine selection chart. Source: Ref. [12].

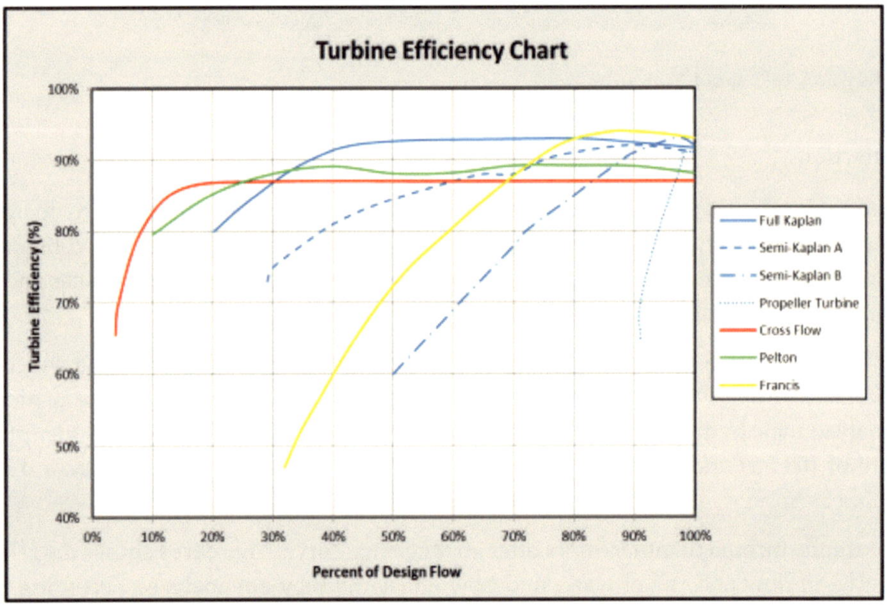

Figure 9. Turbine efficiency chart. Source: Ref. [12].

4. Socioeconomic and environmental aspects

Small hydropower is a key element for sustainable development due to the following reasons [1]:

- Proper utilization of water resources: in small hydro, small creeks, and streams are able to provide and generate energy. You can enjoy local without large storage of water, reducing the social and environmental impacts for the local population.
- Small hydropower is a renewable source of energy: the resource used by SHP, the water, to generate energy is a renewable resource. Therefore, this project is classified as a renewable energy to enjoy the water and generate electricity.
- Small hydro is a cost effective and sustainable source of energy: the SHPs have a simple construction, smaller, and its operating equipment to generate power is low cost compared to large plants. The cost of electricity generation is the free inflation. The period includes the construction and operation is short and the financial return happens quickly.
- Small hydro aids in conserving scarce fossils fuels: the use of water to generate electricity by power plants replaces fossil fuels and petroleum products. If there is the possibility of replacing nonrenewable resources, the SHP is a good choice.
- Low polluting: one of the great contemporary concerns is the relationship of power generation with the environment and reducing the negative impacts. Renewable energy sources reduce GHG emissions and contribute to sustainability. There is a research that puts the SHP as a renewable energy source that reduces GHG emissions and assist in the sustainable development of rural regions. As the SHP does not have large reservoirs and the adaptation of the local population with the project does not suffer many impacts, it is a good choice for electrical projects. The technology of SHP should be harnessed with a form of mitigation of greenhouse gases, along with other renewable forms of generating electricity [1].
- Development of rural and remote areas: there is a deployment potential in remote and mountainous areas for the installation of small power plants. The use of this renewable source of energy in these regions allows the economic and social development.
- Other uses: other benefits are found in regions where small plants are installed, such as irrigation, water supply, tourism, fisheries, and flood prevention.
- The SHP technology is solid and its power house can be built in a few years and has a great life cycle. The civil engineering works, as the dam, can operate for more than a century and require little maintenance. In other mechanical equipment such as turbine, there is the development of research to increase their energy efficiency and reach levels of up to 90% utilization [14].

5. Conclusion

Small hydropower technology is one of the most common technologies used for electricity generation for rural population in both developed and developing countries. Inclusion of the remains of this resource in the energy mixes could lead to sustainable development. Small hydroelectric power plants contribute to meeting the needs of regions where there is no a major technological development and they are able to improve the population's quality of life with the creation of jobs, increase the local economy, and enhancement of the region.

The benefits of the SHP projects fit into the ease of smaller investments and faster build-operate periods. The areas for power generation are smaller, enjoy raw materials, and local labor and the cost of generation compared to other energy projects are also lower. However, the social, political, economic, historical, regulatory, and environmental issues may limit further development of this technology.

Author details

Jacson Hudson Inácio Ferreira[1,*] and José Roberto Camacho[2]

*Address all correspondence to: jacson@iftm.edu.br

1 Federal Institute of Triângulo Mineiro, Brazil

2 Federal University of Uberlândia, Brazil

References

[1] Liu H, Masera D, Esser L, eds. World Small Hydropower Development Report 2013. United Nations Industrial Development Organization, Vienna, Austria; International Center on Small Hydro Power, Hangzhou, China. 2013. Available from http://www.smallhydroworld.org.

[2] Ferreira JHI, Camacho JR, Malagoli JA, Guimarães Júnior SC. Assessment of the potential of small hydropower development in Brazil. Renewable and Sustainable Energy Reviews. 2016;**56**:380-387.

[3] Okot DK. Review of small hydropower technology. Renewable and Sustanaible Energy Reviews. 2013;**26**:515-520.

[4] Kosnik L. The potential for small scale hydropower development in the US. Energy Policy. 2010;**38**:5512-5519.

[5] Gatte MT, Kadhim RA. Hydro Power, Energy Conservation, Dr. Azni Zain Ahmed (Ed.), Intech, 2012, Hilla, Iraq. DOI: 10.5772/52269. Available from: http://www.intechopen.com/books/energy-conservation/hydro-power

[6] Ferreira JHI. A contribution to the study of the estimated hydroelectric potential of small hydraulic plants [dissertation]. Uberlândia – MG, Brazil: 2014. 108 p.

[7] Brazilian Power Plants. Guidelines for studies and small-scale projects Hydraulics. Rio de Janeiro, Brazil, 2000. 458 p. {In Portuguese}.

http://dx.doi.org/10.5772/66532

[8] Forouzbakhsh F, Hosseini SMH, Vakilian M. An approach to the investment analysis of small and medium hydro-power plants. Energy Policy 2007;**35**:1013-1024.

[9] Hosseini SMH, Forouzbakhsh F, Rahimpoor M. Determination of the optimal installation capacity of small hydropower plants through the use of technical, economic and reliability indices. Energy Policy 2005;**33**:1948-1956.

[10] Ogayar, Vidal PG. Cost determination of the electro-mechanical equipment of a small hydropower plant. Renewable Energy 2009;**34**:6-13.

[11] Balat H. A renewable perspective for sustainable energy development in Turkey: the case of small hydropower plants. Renewable and Sustainable Energy Reviews. 2007;**11**(9):2152-2165.

[12] Johnson K, George L, Panter J. Small Hydropower Handbook. Colorado Energy Office 2015. 96 p. Denver, USA.

[13] Cada, GF, Copping AE, Roberts J. Ocean/tidal/stream power: identifying how marine and hydrokinetic devices affect aquatic environments. Hydro Review. 2011;**30**, 3/6 p.

[14] Kong Y, Wang J, Kong Z, Song F, Liu Z, Wei C. Small hydropower in China: the survey and sustainable future. Renewable and Sustainable Energy Reviews 2015;**48**:425-433.

Chapter 3

Design of Zero Head Turbines for Power Generation

Edwin Chica and Ainhoa Rubio-Clemente

Additional information is available at the end of the chapter

http://dx.doi.org/10.5772/66907

Abstract

Failure analysis of the blades of a horizontal axis hydrokinetic turbine of 1 kW is presented. Analysis consisted of the determination of the pressure on the blade surface using Computational Fluid Dynamics, and the calculation of the stress distribution in the blade due to hydrodynamic, inertial and gravitational loads using the finite element methods. The results indicate that the blade undergoes significant vibration and deflection during the operation, and the centrifugal and hydrodynamic loads considerably affect the structural response of the blade; however, the stresses produced in all of the analysed models did not exceed the safe working stresses of the materials used to manufacture the blade. Modal analysis was conducted to calculate first significant natural frequencies. Results were studied in depth against operating frequency of the turbine. After carrying out the modal analysis, harmonic analysis was also done to see the response of the turbine under dynamic loading. It was observed that the turbine is safe in its entire operating range as far as phenomenon of resonance is concerned. Additionally, it was observed that maximum harmonic response of the turbine on the application of dynamic loading is far lesser than its failure limit within the specified operating range.

Keywords: horizontal axis hydrokinetic turbine, tip-speed ratio, power coefficient, blade design, blade load

1. Introduction

In order to generate new and sustainable solutions for the energy crisis and climate change, many alternatives have been studied for increasing the potential of use of renewable energy systems. The widely known alternative energy resources are solar, wind, wave, hydraulic, biomass, and geothermal, among others. However, hydrokinetic energy is one of the renewable energies gaining a great global attention in recent years, especially because it is a stable, continuous, sustainable, and high-energy density renewable resource. Moreover, its costs and development time are small, resulting in a minimal environmental impact. In addition, many countries have a large hydrokinetic potential thanks to their long rivers with stable and high flow rates [1, 2].

Hydrokinetic systems convert the energy of moving water in rivers, oceans, or tidal currents into electricity without the use of a dam or barrage associated to conventional hydropower. Nevertheless, hydrokinetic energy conversion systems are still in early stages of development; therefore, further researches are required [1, 2]. The main challenges linked to the use of hydrokinetic energy are not only to have efficient systems but also to convert energy in a more economical way so that the cost-benefit analysis leads to the growth of this alternative energy generation. For the hydrokinetic energy extraction from water, a turbine submerged in the flowing stream of water is needed, which must be designed for operating conditions of the location where it will be placed to obtain the highest efficiency production. The turbine should be ideally installed at locations that have relatively steady flow throughout the year and are not prone to serious flood events, turbulence, or extended periods of low water level [2–4].

There are several kinds of hydrokinetic turbines of various sizes and different energy capture principles. The major classification of hydrokinetic turbines is related to the rotating axis positions with respect to the water flow: (a) horizontal axis hydrokinetic turbines (HAWTs), where the rotational axis must be oriented in parallel with regard to the water current in order to produce power and (b) vertical axis hydrokinetic turbines (VAWTs), where the rotational axis is perpendicular to the water current direction [4, 5].

Regardless the type of turbine, HAWTs and VAWTs are still at a pre-commercial stage, although a few of them have been developed at a full scale. This fact can be adequate to decentralize the electricity generation in off-grid remote communities, particularly, in developing countries where the kinetic energy conversion may be possible due to the strong currents present in the rivers. In this sense, these ones might be installed in isolated or grid-connected configuration, standing alone, or as a supplement to an existing generation system, floating or fixed to the bottom of the water current. Their construction methods and materials should be appropriate for their local manufacture in these countries [3, 5–7].

The design philosophies of hydrokinetic turbines are similar to that of wind turbines, sharing the same working principles, which consist of converting the hydrokinetic power into mechanical power in the form of rotating blades. In general, many hydrokinetic turbine researches are focused on accurately predicting their efficiency. Various computational models exist, each one with their own strengths and weaknesses, attempting to precisely predict the performance of a hydrokinetic turbine. Being able to numerically predict the hydrokinetic turbine performance offers the possibility to reduce the number of the required experimental tests. Additionally, computational studies are more economical than the costly experiments [5–7].

It is noteworthy that numerous sources claim that HAWTs have a major efficiency per same swept area. Nevertheless, the main advantage of VAWTs is that the blades can have a constant shape along their length and, unlike HAWTs, twisting the blade is not required, as every section of the blade is subjected to the same water speed, allowing an easier design, fabrication, and replication of the blade, which can influence in a cost reduction. This is the reason why this kind of rotor configuration for the design of hydrokinetic turbines is selected. Nonetheless, VAWTs are not as efficient as HAWTs, and they exhibit a very low starting torque, as well as dynamic stability problems [5–7].

The structural design of HAWT blades is also as important as their hydrodynamic design due to the dynamic structural loads a rotor may experience play the major role in determining its lifetime. For this purpose, modal analysis provides information on the dynamic characteristics of structural elements at resonance and, thus, it helps in understanding their detailed dynamic behavior. It is important to note that a good hydrokinetic system should be able to mitigate the unwanted vibratory response, which can only be done by avoiding resonance. Hence, modal or resonance frequency should be computed during the design phase of the blade [6, 7]. Hydrodynamic loads are a major source of dynamic structural behavior and ought to be accurately determined. Under this scenario, this work is aimed at contributing to the better comprehension of the structural static and dynamic behavior of three internal blade configurations using Computational Fluid Dynamic (CFD) simulations and Finite Element Analysis (FEA).

2. Hydrodynamic design of a horizontal axis hydrokinetic turbine

The performance of a hydrokinetic turbine is a main concern in the design stage. It depends on the interaction between the rotor and the water current. The rotor is the most important component in a hydrokinetic turbine, which must be designed for capturing water energy and converting it into rotating mechanical energy. The rotor consists of several blades joined to a common hub [6, 7]. The turbine blades are critical components of the rotor, which use the flow to cause a pressure difference on either side of the hydrofoil profile, creating a resultant force on the blade profile and, subsequently, causing a rotational movement. This rotating movement of the rotor shaft is passed, in most cases, through a gearbox to increase the amount of revolutions of the moving rotor. The output of the gearbox is directly coupled to a generator, which is used to produce electricity and transfer it to the desired location via cables [1, 4, 5, 8].

The geometry and dimension of the rotor are determined by the performance requirements of the hydrokinetic turbine. In general, two fundamental issues must be considered simultaneously in the rotor design process: the hydrodynamic performance and the structural design. The hydrodynamic characteristics of the turbine are influenced by the chosen hydrofoil. Generally, an optimum turbine blade is narrower at the tip and wider near the root. It must be designed to meet structural requirements and it can be manufactured easily. There are also other important design considerations such as the blade materials, noise reduction, condition/health monitoring, blade roots, and hub attachment, among others [3, 4].

The design methods of the main characteristics of the turbine rotor are based on the same incompressible flow techniques used for designing wind turbines [4, 5, 9]. The first step involved in the rotor design is to fix the turbine output power and estimate the blade radius.

For this purpose, it is widely known that the energy in moving water is in the form of kinetic energy, which may be expressed using Eq. (1):

$$E = \tfrac{1}{2} m V_1^2 \tag{1}$$

where m and V_1 correspond to the mass and the velocity of the object, respectively. In a hydrokinetic turbine, there are several units of water masses moving perpendicularly to the plane or area (A) swept out by the blades. This area can be expressed as $A = \pi R^2$, where R is the blade radius. In turn, the power (P) is the rate of energy movement per unit of time or the rate at which energy is generated or consumed per unit time. Mathematically, P is the first derivative of the energy taken with respect to time (dE/dt). This power is developed in the hydrokinetic turbine from the mass movement of water, as indicated in Eq. (2):

$$P = \frac{dE}{dt} = \frac{1}{2}\frac{dm}{dt} V_1^2 \tag{2}$$

The next step is to quantify the amount of water flow associated with dm/dt. Given a unit of time (t), the unit of water masses (m) is moved a distance of L. This can be used to arrive at two important results. First, the first derivative of L with respect to t is dL/dt, which is equal to velocity (V_1). Second, the volume of water passing through A will be AL.

If the density of water (ρ) is known or can be found, then the mass of water moving through the plane A is $m = \rho AL$. The mass movement rate dm/dt is equal to $\rho AdL/dt$ or the expression represented in Eq. (3):

$$\frac{dm}{dt} = \rho A V_1 \tag{3}$$

Substituting Eq. (3) into Eq. (2), Eq. (4) is obtained:

$$P = \frac{1}{2}\rho A V_1^3 \tag{4}$$

Eq. (4) expresses the ideal power in a fluid flow. However, hydrokinetic turbines are limited by blade efficiencies, mechanical losses in transmissions, electrical losses, and the theoretical amount of energy allowed to be extracted from the water current. Because of these losses and inefficiencies, two more variables must be added to Eq. (4). The first variable is η, which is a measure of the efficiency of the gearbox, the electrical inverter, and the generator. This variable takes into account all the friction, slippage, and heat losses associated with the internal mechanical and electrical components. Values of η may greatly differ among the different turbine models. A reasonable and conservative value of η can be around 70% [1, 8].

The second variable is the power coefficient (C_p), which refers to a measure of the blade or hydrofoil efficiency. It includes the hydrofoil shape and the hydrodynamic forces of lift and drag. The C_p expresses the hydrofoil ability to transform the water kinetic energy into mechanical power, which is delivered to the turbine transmission or directly to the electrical generator. A higher C_p is preferred over a lower C_p value. These C_p values may vary with the turbine size.

Placing these two variables, η and C_p, into Eq. (4), the following expression is obtained:

$$P = \frac{1}{2}\rho A V_1^3 C_p \eta \quad (5)$$

Eq. (5) is considered to determine the power of the hydrokinetic turbine. P is the net power derived from water after accounting for losses and inefficiencies. From this equation, it can be seen that the extractable power is dependent on the turbine power coefficient, the density of the flow medium, the swept area of the turbine, and the mean velocity of water (V_1) [8].

If the water density, blade radius, and water velocity are constant, the captured power (P) is proportional to the C_p, which depends on the tip-speed ratio (TSR) (λ) and the pitch angle of the turbine (θ), as in wind turbines. The maximization of C_p is of fundamental importance in order to optimize the extraction of energy from water. It is highlighted that λ is the ratio between the speed of the blade at its tip and the speed of the water current. This ratio has a strong influence on the efficiency of the turbine [4, 5, 8, 10]. This variable may be defined as Eq. (6):

$$\lambda = \frac{R\omega}{V_1} \quad (6)$$

where ω is the rotational speed of the rotor of the hydrokinetic turbine (rad/s). The calculation of the C_p requires the use of the blade element theory. However, numerical approximations have been developed [11]. In the current work, Eq. (7) was used:

$$C_p(\lambda, \theta) = 0.22\left(\frac{116}{\lambda_i} - 0.4\theta - 5\right) e^{\frac{-12.5}{\lambda_i}} \quad (7)$$

$$\frac{1}{\lambda_i} = \frac{1}{\lambda + 0.08\theta} - \frac{0.035}{\theta^3 + 1} \quad (8)$$

With the function defined in Eq. (8), it is possible to evaluate C_p at different values of tip-speed ratio and pitch angle. This leads to the C_p (λ, θ) versus λ characteristics for various values of θ as depicted in **Figure 1**.

In turn, **Figure 2** shows that for every θ, there is a λ, which corresponds to the maximum C_p and, hence, to the maximum efficiency. Note that in **Figure 2**, C_p has a maximum at 0.4382 when θ and λ are equal to 0 and 6.325, respectively. This has two important connotations. On the one hand, achieving a maximum at $\theta = 0$ means that any θ deviation yields lower power capture. On the other hand, maximum conversion efficiency is accomplished at $\lambda = 6.325$.

Before using Eq. (6), V_1 have to be determined or assumed. In this work, V_1 was assumed to be 1.5 m/s; the turbine output power, 1 kW; the drive train efficiency, 70%; and ρ, 997 kg/m^3 (25°C). Now, the radius and, subsequently, the length of the turbine blades may then be calculated. The required blade length resulted to be equal to 0.79 m.

The analysis of the hydrodynamic behavior of hydrokinetic turbine can be done similarly to the simplest model of a wind turbine. Therefore, a simple model, generally attributed to Betz

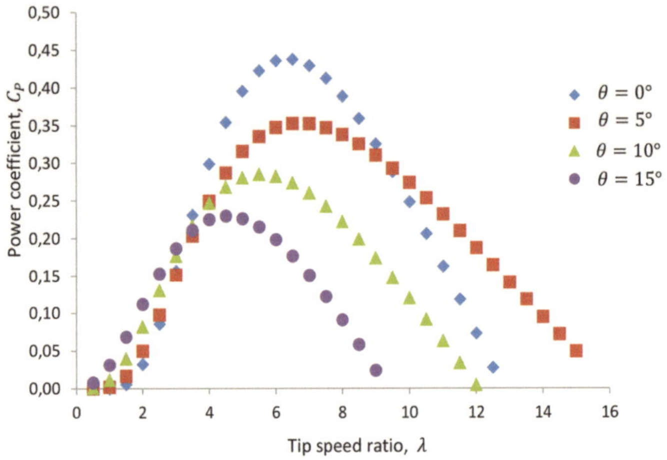

Figure 1. Power coefficient (C_p) as a function of the tip-speed ratio (λ) with the pitch angle (θ) as a parameter.

(1926), can be used to determine the power, the thrust of the water on the ideal rotor, and the effect of the rotor operation on the local water field [4, 5, 9, 12]. The referred model is known as "actuator disk model" where the rotor is replaced by a homogenous disk extracting energy from water. Actuator disk model is based on the assumptions like no frictional drag, homogenous, incompressible, steady-state fluid flow, constant pressure increment or thrust per unit area over the disk, continuity through the disk, and an infinite number of blades. The analysis of the actuator disk model assumes a control volume as illustrated in **Figure 2**.

The analysis of the actuator disk theory assumes a control volume in which the boundaries are the surface walls of a stream tube and two cross sections. In order to analyze this control volume, four stations (1: free stream region, 2: just before the blades, 3: just after the blades, 4: far wake region) need to be considered (**Figure 2**). Applying the conservation of linear momentum equation on both sides of the actuator disk, the net force on the contents of the control volume can be found. That force is equal and opposite to the thrust (I), which is the force of water on the hydrokinetic turbine. I is equal and opposite to the change in momentum of water stream. For steady-state flow, assuming the continuity of mass flow, I can be obtained as expressed in Eq. (9):

$$I = \dot{m}(V_1 - V_4) = \dot{m}_2(V_1 - V_4) = \rho A_2 V_2(V_1 - V_4) \tag{9}$$

I can be also expressed as the sum of the force on each side of the actuator disk, as represented in Eq. (10):

$$I = A_2(P_2 - P_3) \tag{10}$$

Since I is positive, the velocity behind the rotor (V_4) is less than the free stream velocity (V_1). On the other hand, as the flow is frictionless and no work or energy transfer is done, Bernoulli

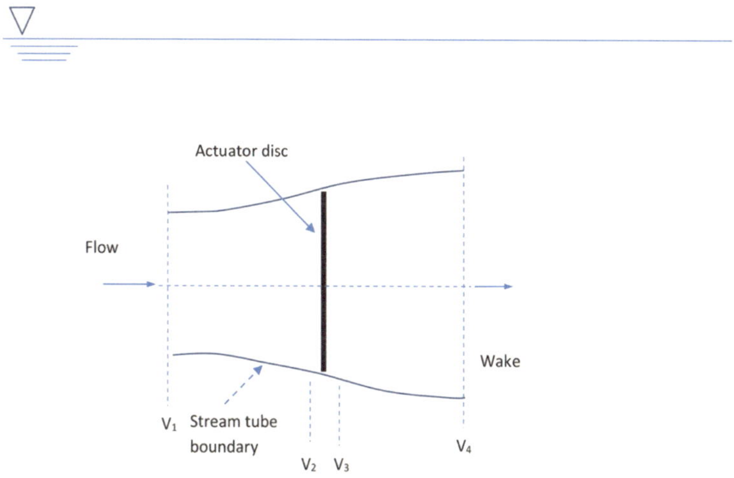

Figure 2. Idealized flow through a hydrokinetic turbine represented by a nonrotating actuator disk.

equation can be applied on both sides of the rotor. If energy conservation using Bernoulli equation is applied between stations 1 and 2 and stations 3 and 4, and they are combined, it is possible to find the pressure decrease (P_2–P_3):

$$(P_2 - P_3) = \frac{1}{2}\rho(V_1^2 - V_4^2) \tag{11}$$

Equating *I* values from Eqs. (9) and (10) and substituting Eq. (11), Eq. (12) can be obtained:

$$V_2 = \frac{V_1 + V_4}{2} \tag{12}$$

The water velocity at the rotor plane, using the simple model, is the average of the upstream and downstream water speeds. If the axial induction factor (*a*) is defined as the fractional decrease in water velocity between the free stream and the rotor plane, then Eq. (13) can be achieved:

$$a = \frac{V_1 - V_2}{V_1} \tag{13}$$

Substituting Eq. (13) into Eq. (12), Eqs. (14) and (15) can be obtained:

$$V_2 = V_1(1 - a) \tag{14}$$

$$V_4 = V_1(1 - 2a) \tag{15}$$

The power captured by a hydrokinetic turbine rotor (*P*) is equal to *I* times the velocity at the rotor plane:

$$P = \rho A_2 V_2 (V_1 - V_4) V_2 \tag{16}$$

Substituting Eqs. (14) and (15) into Eq. (16), Eq. (17) can be achieved:

$$P = 2\rho A_2 V_1^3 a (1 - a)^2 \tag{17}$$

Equating the power values from Eqs. (17) and (5), Eq. (18) is attained:

$$C_p = \frac{4a (1 - a)^2}{\eta} \tag{18}$$

The maximum C_p is determined by taking the derivative of Eq. (18) with respect to a and setting it equal to zero, which yields Eq. (19) when $a = 1/3$ [9]:

$$C_p = \frac{16}{27} = 0.5926 \tag{19}$$

This result indicates that if an ideal rotor is designed and operated so that the water speed at the rotor is 2/3 of the free stream water speed, then it would be operating at the point of maximum power production. This is known as the Betz limit [9]. Thereby, for the hydrokinetic turbine of the best rotor design, extracting more than about 60% of the kinetic water energy is evidently non possible. In the design of the horizontal axis hydrokinetic turbine of 1 kW, the value of $C_p = 0.5926$ was not used. The used value was obtained from Eq. (7) [9]. However, the value of a was determined from Eq. (18) for a C_p equal to *0.4382* and η equal to *100%*. In this case, a was *0.15254*.

In order to make the results more realistic, the effect of wake rotation should be considered. Two assumptions were made to describe this effect. The upstream flow was assumed to be entirely axial, while the downstream flow was assumed to have rotation at an angular speed (Ω). By considering the tangential flow behind the rotor, the angular induction factor (a') was introduced [9]:

$$a' = \frac{\Omega}{2\omega} \tag{20}$$

where Ω and ω refer to the induced tangential angular velocity of the flow and the angular velocity of the rotor, respectively. Equations for the torque and the thrust force can be obtained by considering the flow through an annular cross section of radius r with area $2\pi r dr$. Thus, the experienced thrust force and torque can be expressed as Eqs. (21) and (22), respectively:

$$dI = 4a'(1 + a')\frac{1}{2}\rho\,\omega^2 2\pi\, r^2\, dr \tag{21}$$

$$dT = 4a'(1 - a)\frac{1}{2}\rho\, V_1^2\,\omega\, r^2\, 2\pi \mathrm{r dr} \tag{22}$$

The thrust on an annular cross section can be also determined by the following expression that uses the axial induction factor, a:

$$dI = 4a(1 - a)\frac{1}{2}\rho V_1^2 2\pi r dr \tag{23}$$

In order to obtain a relationship between the axial induction factor and the angular induction factor, Eqs. (21) and (23) can be equated, giving Eq. (24):

$$\frac{a(1 - a)}{a'(1 + a')} = \lambda_r^2 \tag{24}$$

where the local tip-speed ratio (λ_r) is defined as Eq. (25):

$$\lambda_r = \frac{\omega r}{V_1} = \lambda\frac{r}{R} \tag{25}$$

On the other hand, the blade element theory is useful to derive expressions of developed torque and the axial thrust force experienced by the turbine. This theory is based on the analysis of the hydrodynamic force applied to a radial blade element of infinitesimal length. **Figure 3** illustrates a transversal cut of the blade element. In the figure, the hydrodynamic force acting on the blade element is also depicted. The blade element moves in the water flow at a relative speed (V_{Rel}). The flow around the blade starts at station 2 in **Figure 2** and ends at station 3. At the inlet to the blade, the flow is not rotating, and at the exit from the blade, the flow rotates at a rotational speed (Ω.). That is, over the blade row, wake rotation has been introduced. The average rotational flow over the blade due to wake rotation is, therefore, Ω/2. The blade is rotating with speed ω. The average tangential velocity that the blade experiences is, thus, $\omega r + \frac{1}{2}\Omega r$. This is shown in **Figure 3**.

A blade can be divided into N elements as shown in **Figure 3**. Each of the blade elements will experience a slightly different flow as they have a different rotational speed (ωr) and a different chord length (c). The water flow establishes a differential pressure around the blade element, which results in a force perpendicular to the local water movement direction, the so-called lift force (dL). Additionally, a drag force (dD) is done in the flow direction. The drag force is due to both viscous friction forces at the surface of the hydrofoil and the unequal pressure on the hydrofoil surface facing toward and away from the oncoming flow [9]. dL and dD can be found from the definition of the lift (C_L) and drag (C_D) coefficients as follows:

$$dL = C_L\frac{1}{2}\rho V_{Rel}^2 c dr \tag{26}$$

$$dD = C_D\frac{1}{2}\rho V_{Rel}^2 c dr \tag{27}$$

Both lift and drag coefficients are functions of the type of hydrofoil used for the blade and the incidence angle (α), defined as the angle that the flow makes with the chord. As observed in **Figure 3**, it is possible to obtain the distribution of the angle of the relative water velocity (ϕ) which consists of the pitch angle (θ), twist angle (β), and angle of attack (α) on one of the blade sections (hydrofoil) by employing blade element momentum theory, as expressed in Eq. (28):

$$\phi = \alpha + \theta + \beta \tag{28}$$

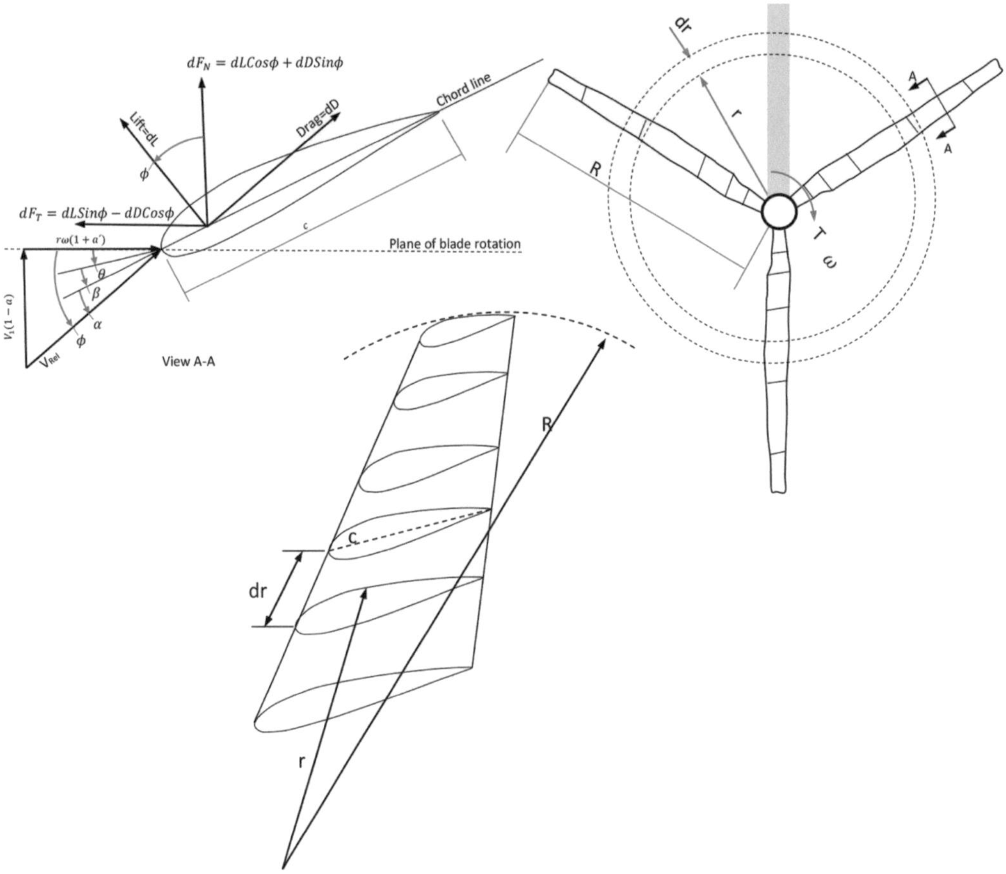

Figure 3. The blade element model.

where ϕ is the angle between the local flow direction and the rotor plane, and β is the twist angle. However, C_p has a maximum when θ is equal to 0; therefore, Eq. (28) can be rewritten as Eq. (29):

$$\phi = \alpha + \beta \tag{29}$$

Examining **Figure 3**:

$$\text{Tan}\phi = \frac{V_1(1-a)}{\omega r(1+a')} \tag{30}$$

Substituting Eq. (25) into Eq. (30), the expression for $Tan\phi$ can be further simplified to Eq. (31):

$$\text{Tan}\phi = \frac{(1-a)}{\lambda_r(1+a')} \tag{31}$$

http://dx.doi.org/10.5772/66907

From **Figure 3**, the relative water velocity can be expressed as a function of the water velocity and the rotational speed as expressed by Eqs. (32) and (33), respectively:

$$V_{Rel} = \frac{V_1(1 - a)}{\mathrm{Sin}\phi} \tag{32}$$

$$V_{Rel} = \frac{\omega r(1 + a')}{\mathrm{Cos}\phi} \tag{33}$$

The lift and drag forces can be resolved into a normal force (dF_N) to the plane of rotation (this contributes to thrust) and a tangential force (dF_T) to the circle swept by the rotor. This is the force creating useful torque. Therefore, if there are B number of blades, the torque (dT) on the rotor of a hydrokinetic turbine is simply the tangential force multiplied by the radius:

$$d F_N = \mathrm{dLCos}\phi + \mathrm{dDSin}\phi = (C_L \mathrm{Cos}\phi + C_D \mathrm{Sin}\phi)\frac{1}{2}\rho V_{Rel}{}^2 cdr \tag{34}$$

$$d F_T = \mathrm{dLSin}\phi - \mathrm{dDCos}\phi = (C_L \mathrm{Sin}\phi - C_D \mathrm{Cos}\phi)\frac{1}{2}\rho V_{Rel}{}^2 cdr \tag{35}$$

$$dI = B\frac{1}{2}\rho V_{Rel}{}^2(C_L \mathrm{Cos}\phi + C_D \mathrm{Sin}\phi)cdr \tag{36}$$

$$dT = B\frac{1}{2}\rho V_{Rel}{}^2(C_L \mathrm{Sin}\phi - C_D \mathrm{Cos}\phi)\mathrm{crdr} \tag{37}$$

The power developed by the rotor (dP) can be found as the product of the torque and the angular velocity, as expressed in Eq. (38):

$$dP = B\omega\frac{1}{2}\rho V_{Rel}{}^2(C_L \mathrm{Sin}\phi - C_D \mathrm{Cos}\phi)\mathrm{crdr} \tag{38}$$

In order to obtain the chord length (c), Eqs. (37) and (22) can be equated, resulting in Eq. (39):

$$c = \frac{8a'\,\omega\, r^2\, \pi\, \mathrm{Sin}^2\, \phi}{(C_L \mathrm{Sin}\phi - C_D \mathrm{Cos}\phi)B(1 - a)} \tag{39}$$

The number of blades (B) will affect the power output from the turbine. The optimum number of blades for a turbine depends on the purpose of the turbine. Turbines for generating electricity need to operate at high speeds, but they do not require much torque. These turbines generally have two or three blades, since this gives enough torque without adding the extra weight that can slow the turbine down. Rotors with odd numbers of blades, and at least three, are more stable. The advantage of having one or two blades in the rotor is the possible savings in production costs and weight. Nevertheless, the use of few blades requires a higher rotational speed or a larger chord length to yield the same energy output. A three-blade design also decreases the fluctuating loads from inertia variation [4, 5, 13–15].

In order to apply Eq. (39), it is necessary to introduce a corrective factor (Fc) for improving the structural strength of the blade. Fc was equal to 3.8. Thus, the values of each chord length calculated by Eq. (39) were multiplied by this factor.

On the other hand, in the design of a blade, the most essential aspect is to choose a good profile. A blade profile is typically similar to the wing of an aircraft. The coordinates of hydrofoils are normally given as a function of the chord length (x/c, y/c). These coordinates represent the profile, shape, and thickness of a blade at a particular section along the blade. In the blade design, an S822 hydrofoil profile was used [13, 14]. This hydrofoil profile was developed under a joint effort between the National Renewable Energy Laboratory and Airfoils, Inc. It was specially tailored for using on small horizontal axis turbines. The hydrofoil has several advantages over those hydrofoils traditionally used on aircraft. First, the hydrofoils have a reduced roughness sensitivity for improving the energy capture under dirty blade conditions, owing to the accumulation of insect debris. Second, the increased section thickness of the root (S822, 16%) and tip hydrofoil allows for a lower blade weight, lower cost, increased stiffness, and improved fatigue resistance.

Before using Eq. (39), B, C_L, C_D, a, a', and ϕ must be determined. In this work, B was equal to three. The reasons for that selection were mainly the higher efficiency and smoother output torque than two-bladed turbines. In general, three blades are used for the turbine system to keep the dynamic balance and minimize the fatigue effect. The lift and drag coefficients can be obtained from wind tunnel test data for the type of the used airfoil. The values of C_L (0.8) and C_D (0.009) were used [1, 13, 14]. α, a variable in Eq. (29), may also be found from test data, which in this case was 5°. The value of a was determined from Eq. (18) for a C_p and η equal to 0.4382 and 70%, respectively. In this case, a was 0.15254 and ϕ can be worked out using Eq. (31). However, λ_r must be first determined using Eq. (25).

Once B, C_L, C_D, a, a', and ϕ have been found, c and β can be calculated for any value of r along the blade. In order to simplify, the whole blade length (0.79 m) was divided into 10 equal segments, separating each section by a distance of 0.079 m. The sections were represented from r_1 to r_{10}, where r_1 was the section nearest to the root and r_{10} was the tip of the blade. An Excel spreadsheet shown in **Table 1** was used to obtain the values of c and β for each section. It is highlighted that the value of c increases from the tip toward the root. β was larger near the root and smaller at the tip, where local speeds are lower and higher, respectively. The conditions of the design produced an angular velocity equal to 114.68 rpm.

Once the chord length distribution and the twist distribution along the blade length have been found for a certain tip-speed ratio at which the power coefficient of the rotor is maximum, the next step is to multiply the value of c at every section by the non-dimensional coordinates of the profile S822. The values of x and y coordinates of the profile for each section

Predefined design parameters							
Turbine output power, P (w)	Fluid density, ρ, (kg/m^3)	Water velocity V_1, (m/s)	Power coefficient C_p	Efficiency η	Blade number B	Lift coefficient, C_L	Drag coefficient, C_D
1000	997	1.5	0.4382	0.7	3	0.8	0.009
Tip-speed ratio, λ	Angle of attack, α	Axial induction factor, a Eq. (18)			Rotational speed, ω (rad/s) Eq. (6)	Rotor radius, R(m) Eq. (5)	Δr
6.325	5°	1.291	0.153	0.556	12.099	0.79	0.079

Computed parameters							
Station no. (1 root, 10 tip)	Blade element, r (m)	Local tip-speed ratio, λr Eq. (25)	Angular induction factor, a' Eq. (24)	Relative velocity angle, ϕ (rad). Eq. (31)	Relative velocity angle, ϕ (deg)	Blade twist, β. Eq. (29)	Chord length, c (m). Eq. (39)
1	0.079	0.633	0.257	0.817	46.83	41.83	0.117
2	0.158	1.265	0.075	0.557	31.93	26.93	0.100
3	0.237	1.898	0.035	0.407	23.35	18.35	0.078
4	0.316	2.530	0.020	0.317	18.18	13.18	0.063
5	0.395	3.163	0.013	0.259	14.82	9.82	0.053
6	0.474	3.795	0.009	0.218	12.48	7.48	0.045
7	0.553	4.428	0.007	0.188	10.77	5.77	0.039
8	0.632	5.060	0.005	0.165	9.46	4.46	0.035
9	0.711	5.693	0.004	0.147	8.43	3.43	0.032
10	0.790	6.325	0.003	0.133	7.61	2.61	0.029

Table 1. Excel spreadsheet for determining the section chord length and twist angle.

were exported to parametric 3D design software by using SolidWorks [13, 14]. Therefore, the cross sections of the blade were created at different distances from the root to the tip, and using the Loft command, a 3D model of the whole blade was produced. The resulting image is shown in **Figure 4**.

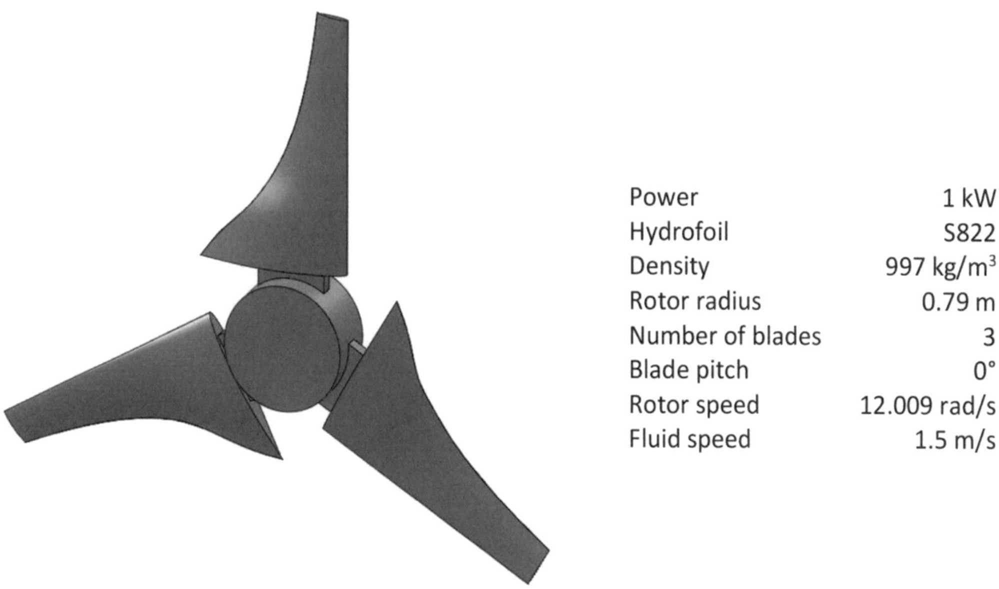

Figure 4. Specifications of the turbine used for simulation studies.

3. Structural design of a horizontal axis hydrokinetic turbine

The rotor hydrodynamic design was validated with a detailed three-dimensional CFD, performed in ANSYs Fluent, and FEA, performed in ANSYs Workbench. The fluid domain was coupled with the structural domain through one-way coupling, and a fluid-structure interaction analysis was carried out to find the effect of the blade geometry and the operating conditions on the stresses developed in the blades.

Based on the characteristic of cyclic symmetry for the structure of the hydrokinetic turbine rotor, a rotor sector conformed by a blade and an angle of 120° of the hub was used to run the simulation. Accordingly, a rotationally periodic boundary condition was applied across the back and bottom surface of the computational domain shown in **Figure 5**, which was created using SolidWorks and meshed in ANSYs. The used mesh was an unstructured mesh with very fine prisms layer near the turbine wall. All dimensions of the boundary were given in terms of radius of the turbine. Since the studied turbine possesses 120° periodicity, only one blade was modeled. **Figure 2** shows the location of the turbine within the computational domain. The turbine rotational plane was located 2R away from the inlet, and the fluid domain extended 5R behind the turbine rotational plane to capture the near and far wake effects. The study assumed steady and incompressible flow. The velocity inlet boundary condition was applied on the left surface of the domain with uniform axial (free stream) velocity of 1.5 m/s. A pressure outlet boundary condition was provided on the right surface with zero gauge pressure. Realizable k-e turbulence model was chosen. The turbulence intensity of 5% was applied. The

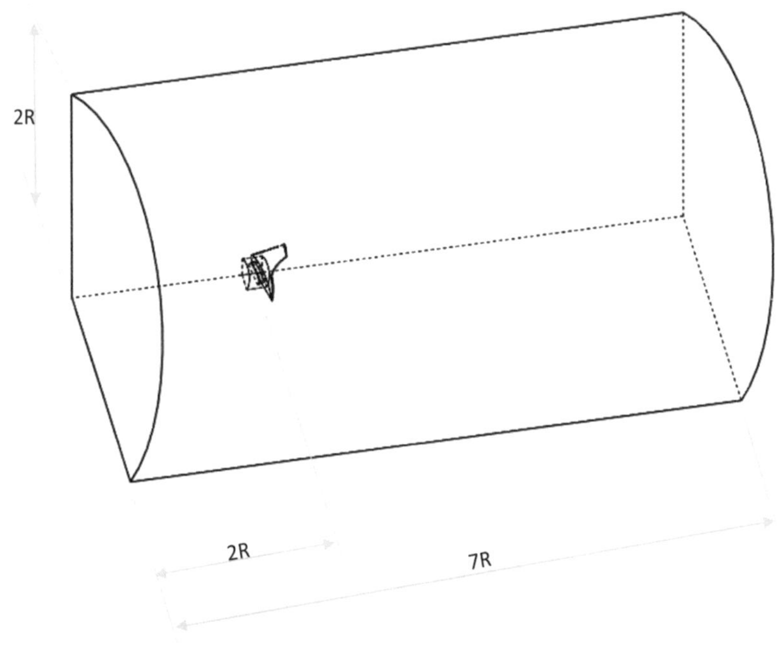

Figure 5. Computational domain used for the CFD analysis.

no-slip condition was applied on the turbine blade and hub, i.e., the relative velocity of the surface blade and hub was set to zero.

A grid independence study was carried out to study the effect of the number of elements on the CFD analysis. The CFD solutions were deemed to have converged when the scaled residuals for all the solution values had dropped by at least six orders of magnitude. The moment on blade was monitored to ensure convergence of the solutions. The CFD simulations provided the flow field within the computational domain and also the blade pressure distribution, shown in **Figure 6**. It can be observed that the pressure on the lower side is relatively higher than the pressure on the upper side, as expected. This creates a net upward force, which produces a lift. Spatial integration of the pressure distribution yielded extreme load distributions, specifically the shearing forces and the bending moments, over the blade.

The simulation allowed computing the torque on the blade (29.366 Nm), which was subsequently multiplied by the blade total number (3) and the angular velocity (12.009 rad/s). Thus, the turbine power output was calculated (1057.969 W). This value was compared to the design power (1000 W). The good agreement between the numerical and theoretical turbine power output enabled to validate the design of the rotor.

The blades of a hydrokinetic turbine must be analyzed for hydrodynamic, centrifugal, and gravitational loads experienced during their operating life. The hydrodynamic loads, i.e., the extreme flap and edgewise loads that the blade would likely encounter over its lifetime, were associated with extreme flow conditions and were determined using CFD software. The centrifugal load was located at the rotation center of the rotor (rotation velocity of 12.009 rad/s), and the gravity was defined in the Static Structural module of ANSYs.

Fluid-structure interaction system was created to perform static analysis by using pressure loads on the blade from Fluent software, transferred them by mapping algorithm, and projected on blade surface geometry in the Static Structural module, following the global

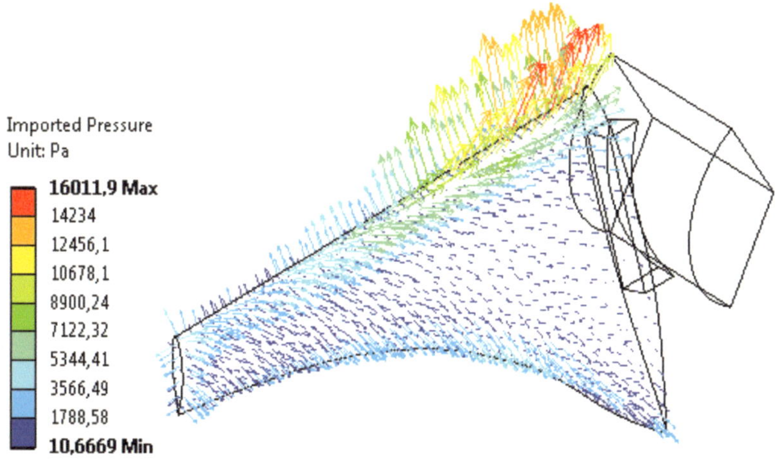

Figure 6. Blade pressure distributions for the extreme operating conditions.

coordinate system (**Figure 4**). That is, the position of the blade geometry in Fluent module and Static Structural module must be the same one with reference to the global coordinate system. **Figure 6** shows the "solution," e.g., the pressure loads on the blade, obtained from the Fluent module, which are connected to the "Setup" of Static Structural module and used as mechanical loads for the blade static analysis (**Figure 7**).

ANSYs Composite PrepPost (ACP) was used for the analysis of layered composite structures of the blade. ACP has a pre- and post-processing mode. In the pre-processing mode, all composite definitions can be created and mapped to the geometry (FE mesh). These composite definitions are transferred to the FE model (Static Structural module) and the solver input file. In the post-processing mode, after the solution is completed and the result file(s) are imported, post-processing results (failure, safety, strains, and stresses) can be evaluated and visualized.

Various blade structural configurations, such as solid blade (Geometry A, **Figure 8A**), face sheet with core (Geometry B, **Figure 8B**), and face sheet with shear webs and stations (Geometry C, **Figure 8C**), were investigated in order to know the global behavior of the blade in terms of its geometry, tip deflection, and stress levels.

The geometry B consisted of two faces or skins (suction and pressure sides) which provided the pressure distribution required for the blade rotation, and thus its cross section was in the form of specific hydrofoils due to hydrodynamic considerations associated with efficiency of the blade in capturing water energy. In addition to the two faces, it had a core made of

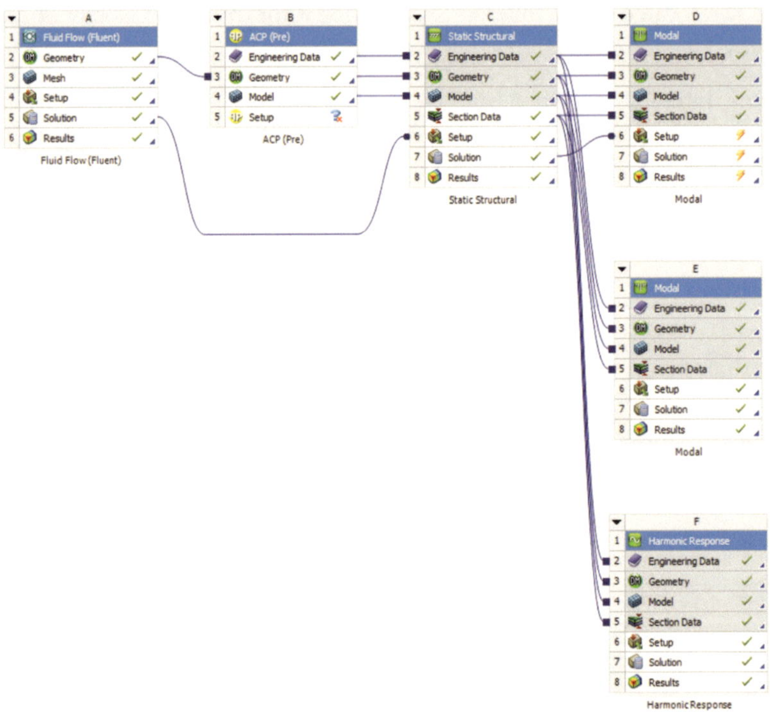

Figure 7. Schematic project of fluid-structure integration (FSI) in ANSYs Workbench.

http://dx.doi.org/10.5772/66907

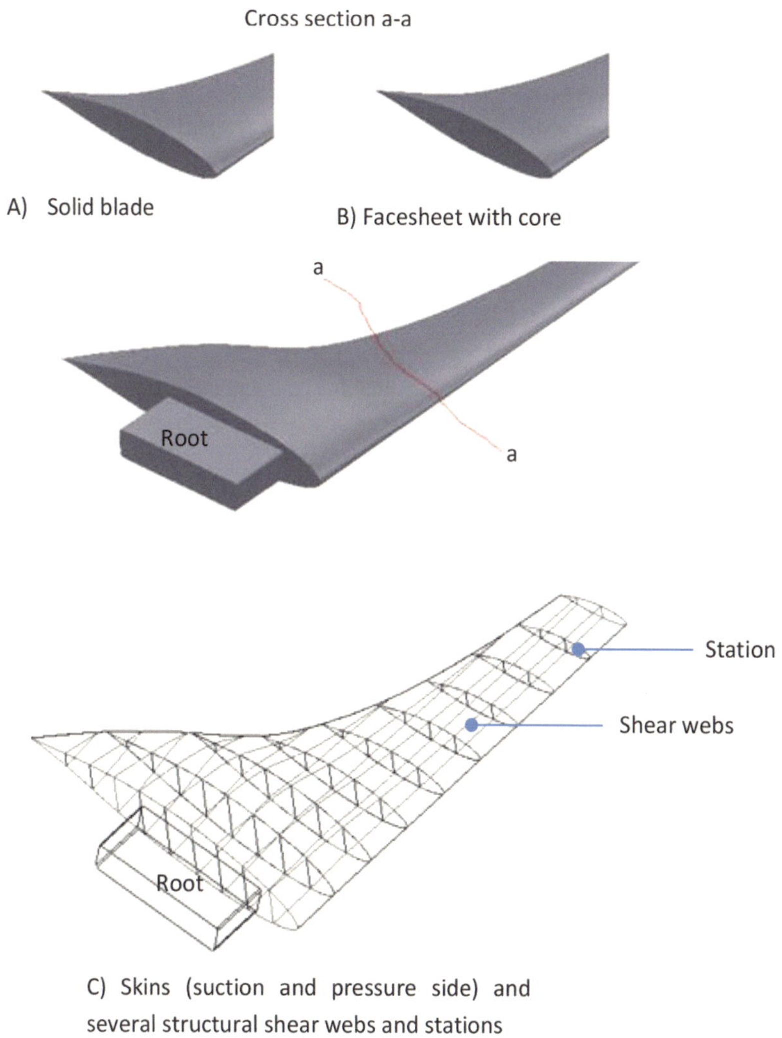

Figure 8. Structural models of the blade. (a) Geometry A, (b) Geometry B, and (c) Geometry C.

a material of low density. This core considerably contributed to the hydrodynamic performance of the blade. The geometry C consisted of two faces or skins (suction and pressure sides), structural shear webs, and several stations. The position, amount, and shape of the shear webs were important in the blade behavior. The shear webs and stations were located inside the shell formed by the faces, which were mainly responsible to support different load cases. The stations or transversal spars had a thickness of 2 mm and were connected to the webs. Since the webs were sheets of 2 mm thickness, they were easy to adjust to the stations to the attack angle of each hydrodynamic profile. The geometrical configurations of the blade were also modeled in SolidWorks. The models presented in **Figure 8** include the 0.79 m long hydrodynamic blade. As it can be observed, this segment starts at a hub, with a rectangular cross section of height 0.035 m and width 0.170 m connected to the first hydrodynamic profile.

The blade model was opened within ACP to define the material types, fiber orientations, and the construction of the layers for each region of the blade (geometries B and C). Since the properties of the composite were different in all of the three perpendicular directions (orthotropic), the model was sectioned into 22 different parts (2 skins, 10 stations, 9 shear webs, 1 root or hub), each one with their own material coordinate system. The skins, shear webs, and stations were modeled as a double laminated ply with a total thickness of 2 mm; each ply would be 1 mm thick with different orientations.

The orthotropic material property descriptions, material thicknesses, and fiber orientations were entered into a material data folder in ACP, where "materials," "fabrics," and "stack-ups" were stored. From the Workbench project page, the loads imported from the CFD analysis, the boundary conditions, and the specific analysis options were applied to the model, and the analysis was executed. Results may be post-processed in mechanical or in ACP to see the stress and strain results in detail on a layer-by-layer basis.

Static numerical models for each geometry of the blade were created: (a) a solid blade (**Figure 8A**) made of nylon; (b) a shell-type blade with skins and a core (**Figure 8B**), whose skins were made of glass fiber (type E), the root of aluminum 6061-T6, and the core of balsa wood; and (c) a shell-type blade (with skins, webs, and stations, see **Figure 8C**), where the skins, the webs, and the stations were made of glass fiber (type E) and the root of aluminum 6061-T6. The mechanical properties of the materials used in the simulation are listed in **Table 2** [10, 16].

The unstructured tetrahedral mesh of the solid blades for the geometry A converged with 12,896 elements and 15,990 nodes. On the other hand, the structure of the blade for the geometry B was meshed with a mixed mesh of tetrahedral elements (root and core) and hexahedral elements (skins). This mesh converged with 13,053 elements and 15,842 nodes. The geometry C was also meshed with a mixed mesh of tetrahedral elements (root) and hexahedral elements (skins, webs, and stations). This mesh converged with 10,308 elements and 10,419 nodes. The thickness of the skins, weds, and stations was 2 mm.

Symmetrical and identical composite lay-up was assumed for skins, shear webs, and stations to improve convergence and achieve a balanced structural design. Two types of ply orientation (0/90 and ±45) for the skins, webs, and stations were chosen for the geometries B and C based on practical selections of ply orientation in industry [16–18].

All the blade models were treated as a cantilevered beam with all the degrees of freedom fixed at the hub. Centrifugal, gravitational, and hydrodynamic forces were used for structural analysis. The von Mises stress and displacements were calculated for different models.

The results for maximum von Mises stresses and displacement over the complete geometry of the blade are presented in **Table 3**. None of the stresses shown in any region of the blade reached the limit values concerning the mechanical strength of the material for yielding. Stresses were maximum at the blade root for all the geometries.

Static structural analysis of geometries B and C shows that the ply orientation of fiber at ±45 was structurally superior in comparison with the ply orientation of fiber at 0/90. The results

http://dx.doi.org/10.5772/66907

Material	Young's modulus (GPa)			Shear modulus (GPa)			Density (kg/m³)	Poisson's ratio			Tensile strength ultimate/yield (MPa)		
	E_1	E_2	E_3	G_{12}	G_{13}	G_{23}	ϱ	ν_{12}	ν_{12}	ν_{12}	x	y	z
Glass fiber (type E)	45	10	10	5	3.8	5	2500	0.30	0.40	0.30	1100	35	35
Balsa wood	3						160	0.29			20		
Aluminum (6061-T6)	69						2700	0.33			310/276		
Nylon	8.3						1400	0.28			139.04		

Table 2. Mechanical properties of the blades [10, 16].

Geometry	Part	Material	Von Mises stress (Mpa)		Displacement results (mm)		Weight (kg)
A	All the blade	Nylon	2.3228		2.1589		8.2628
B	Composite structures of the blade				Ply orientation		
			[0/90]	[±45]	[0/90]	[±45]	
	Skins (suction and pressure side)	Glass fiber (type E)					
	Core	Balsa wood	9.9773	9.0558	0.87592	1.2126	11.0470
	Root	Aluminum (6061–76)					
C	Skins (suction and pressure side)	Glass fiber (type E)					
	Shear webs and stations	Glass fiber (type E)	43.1840	33.3000	1.7754	2.4785	11.3010
	Root	Aluminum (6061–76)					

Table 3. Structural simulation results of the blade.

summarized in **Table 2** show that the blade made up of glass fiber did not fail structurally, due to the stresses produced in all the parts of the analyzed models with glass fiber composite material did not exceed the safe working stresses of this material. Additionally, after the structural design of geometry B it was confirmed that the failure in the core of balsa wood did not occurred through von Mises' criterion. **Figure 9** shows the variation of von Mises stress, displacement, and weight with respect to the analyzed geometries.

On the other hand, the main cause in hydrokinetic turbine blade failure is the high fatigue cycle. Fatigue failure is related to repeated cycling of the load on a structural member. Generally, the fatigue life of a structural member, i.e., the number of load cycles it can survive, is determined by the magnitude of the stress cycles. The exact relation between the magnitude of the stress and the fatigue life depends on the material properties of the structural member. In general, higher stresses lead to a shorter fatigue life. For some materials, fatigue only occurs if stresses exceed a minimum level, while for other materials there is no minimal stress level [4, 19–23].

Fatigue failure occurs when a crack is initiated, which results after a number of load cycles, leading to the concentration of high stress at a specific point of the material, known as point with relatively rough or soft surfaces or with sharp geometrical discontinuities. This crack advances slowly through the material surface up to a certain point from which it is accelerated until the final failure occurs. It is important to note that a high-cycle fatigue corresponds to failure after a relatively large number of load cycles, resulting in stress levels well below the generated material strength, exhibiting an elastic deformation. Furthermore, it is noteworthy that a structural member failure may not be attributed to the excess of load but to the repeated cycling loading [4, 5, 19–23].

In principle, there are two ways in which the failure of turbine blades due to fatigue problems can be eliminated: (a) when an appropriate structural design is performed and (b) when the cycling of the load is presented. The correct design of a structural member can usually

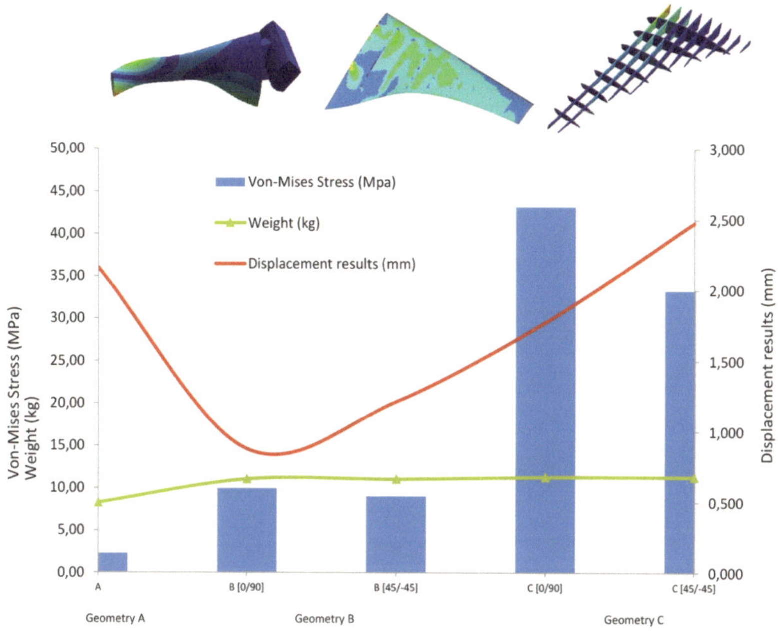

Figure 9. Von Mises stress and displacement on the blade.

eliminate or dramatically reduce fatigue problems. Nevertheless, for a turbine blade, this is not always possible. The design of a blade is usually constrained by hydrodynamic properties, weight, rotor length, etc., which can make the elimination of fatigue problems through design modifications very difficult if not impossible. On the other hand, the prevention of cyclic loading in hydrokinetic turbines is virtually impossible in practice [19, 20].

Modal analysis can be a powerful tool to assist in the identification and elimination of fatigue problems. The most obvious use of modal analysis is the determination of the natural frequencies of the turbine blades. The knowledge of these frequencies can be very useful in avoiding excessive excitations and thereby reducing the risk of fatigue failure. Computer-generated models of the turbine blades can be very useful to investigate the turbine and turbine blade properties under running conditions. Therefore, the modal analysis of the blade was performed using also ANSYs Workbench to extract the natural frequencies and the dominant modes using block Lanczos method [19–21].

For the modal analysis of the blade, loads and constraints were applied as in the structural analysis. First, six significant natural frequencies were calculated for both the cases; i.e., with press stress on and with press stress off. Pre-stressed analysis was performed by doing a static analysis before the modal analysis so that all the stresses and displacements, which are created by the centrifugal, hydrodynamic, and gravitational loads, are included into the modal analysis. In static analysis, rotational speed (in rad/s) was applied.

Results from modal analysis were not evidenced to be highly affected when loads were applied. The comparison of both the results is given in **Table 4**. It can be observed that both

Geometry	Mode no.	Natural frequencies with press stress effect (Hz)	Natural frequencies with press stress effect (Hz)
A	1	59.413	59.344
	2	169.08	169.01
	3	267.07	267.05
	4	378.46	378.38
	5	467.46	467.32
	6	645.85	645.79
B (±45)		98.874	98.828
		312.790	312.750
		399.340	399.320
		761.600	761.560
		817.630	817.540
		1105.400	1105.400
B (0/90)		117.200	117.160
		353.670	353.660
		441.970	441.950
		661.120	661.010
		889.220	889.160
		1055.500	1055.400
C (±45)		70.366	70.302
		199.530	199.490
		264.710	264.620
		345.760	345.700
		443.360	443.330
		489.900	489.870
C (0/90)		83.066	83.010
		215.660	215.590
		299.160	299.100
		400.100	400.040
		462.320	462.150
		510.930	510.900

Table 4. Modal analysis results.

http://dx.doi.org/10.5772/66907

of the results are similar. The resulting mode shapes evidenced for the blades were flexural, axial, or edge bending and torsional modes. The rotational speed of the hydrokinetic turbine designed was 114.68 rpm, so the rotational frequency was 1.911 Hz, resulting from 114.68/60. There were three blades in total in the designed hydrokinetic turbine; therefore, the triple rotational frequency was 5.733 Hz. This rotational frequency and triple rotational frequency are both far smaller than the first six frequencies of the five analyzed models. Thus, the designed hydrokinetic turbine rotors cannot be destroyed by resonance, which satisfies the design demand. As it is widely known, it must be noted that the phenomenon of resonance occurs if the natural frequency of structural system matches the frequency of dynamic loading, which may result in failure of structure. Mode shapes obtained through FEA are presented in **Figure 10**. These mode shapes were computed for a rotational speed of 1.911 Hz. The mode shape contour plots are presented in the form of sum of deformations in the three orthogonal directions in **Figure 10**. It may be observed that the nature of these mode shapes remained largely unaffected by the rotational speed.

Changes in the blade natural frequencies with respect to the machine speed were plotted on a Campbell diagram (**Figure 11**). Critical speeds occur if the excitation frequencies coincide with the natural frequency of the rotating structure. The excitation frequency itself may be

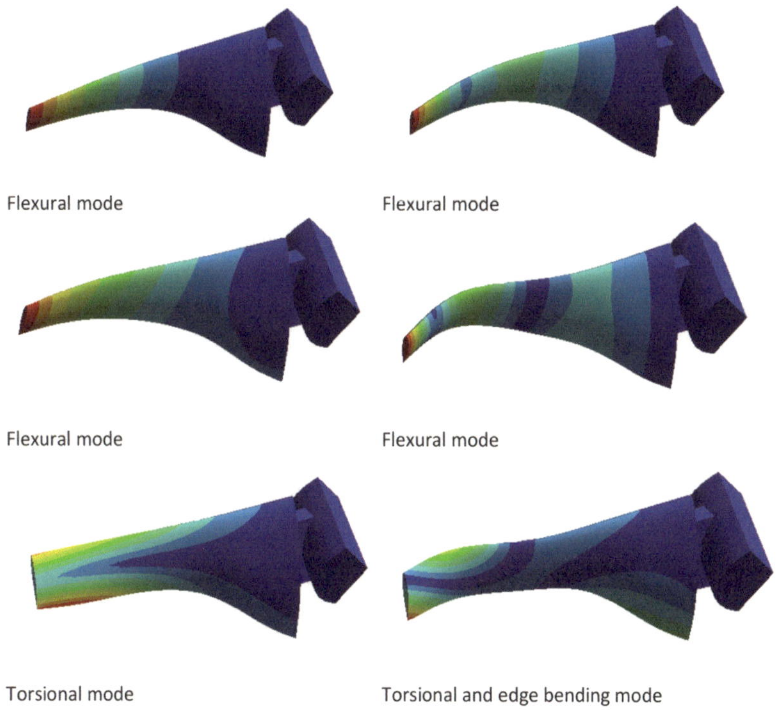

Figure 10. Mode shapes of the turbine blade.

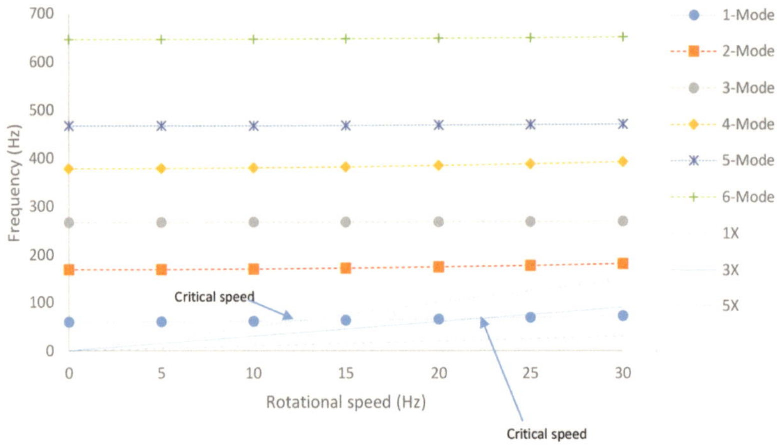

Figure 11. Campbell diagram of the turbine blade for geometry A.

synchronous to the rotational speed (1X) or multiples of it (2X, 3X, 4X, 5X, etc.). These excitation frequencies were represented as straight lines in the Campbell diagram. The intersections of these lines with the natural frequency curves indicated the critical speed. Using this procedure, critical speeds for different excitation frequencies were computed. The critical speeds were 22.407 Hz and 12.612 Hz for 3X and 5X the critical speed, respectively.

To finally calculate the dynamic response of the blade, the mode superposition harmonic analysis was applied. As it is widely known, it must be highlighted that harmonic response analysis is a technique used to determine the steady-state response of a linear structure to loads that vary sinusoidally, i.e., harmonically, with time [20–22]. Harmonic response analysis gives the ability to predict the sustained dynamic behavior of structures; enabling, subsequently, to verify whether or not the designs will successfully overcome resonance, fatigue, and other harmful effects of forced vibrations. Harmonic analysis uses the natural frequencies and mode shapes from the modal analysis. In this case, the frequency spectrum was set between 0 and 700 Hz with 100 steps for the geometry A.

It is highlighted that harmonic analyses require cyclic load data for the analysis. This load data were previously gathered from the hydrodynamic analyses of the turbine carried out by using CFD and static analysis. Therefore, the forces considered in this case were the blade self-weight and the maximum pressure due to the hydrodynamic load. A critical stress region was found from the stress contour obtained by performing harmonic analysis. The stress contours were plotted for the first and second natural frequencies (59.413 Hz and 169.08 Hz, respectively). The maximum stresses observed at these frequencies were 96.147 kPa and 3.015 MPa, respectively. The region of critical stress resulted to be near the root for natural frequencies of 59.4313 and near the tip for frequencies of 169.08 Hz. The von Mises stress variations with frequency at these critical nodes are plotted in **Figure 12**. As observed from these results, the stress in the material was well below the endurance limit. Thus, the structure may be considered safe for the presented loads.

Amplitude versus frequency was plotted using the post-processing option in ANSYs. The frequency response function (FRF) for this analysis is shown in **Figure 13**. These graphics are

http://dx.doi.org/10.5772/66907

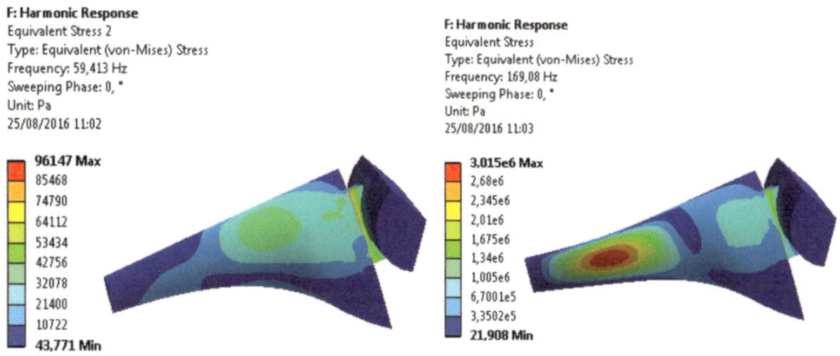

Figure 12. Stress contour (in MPa) at the natural frequency corresponding to 1F mode (59.413 Hz) and 2F mode (169.08 Hz).

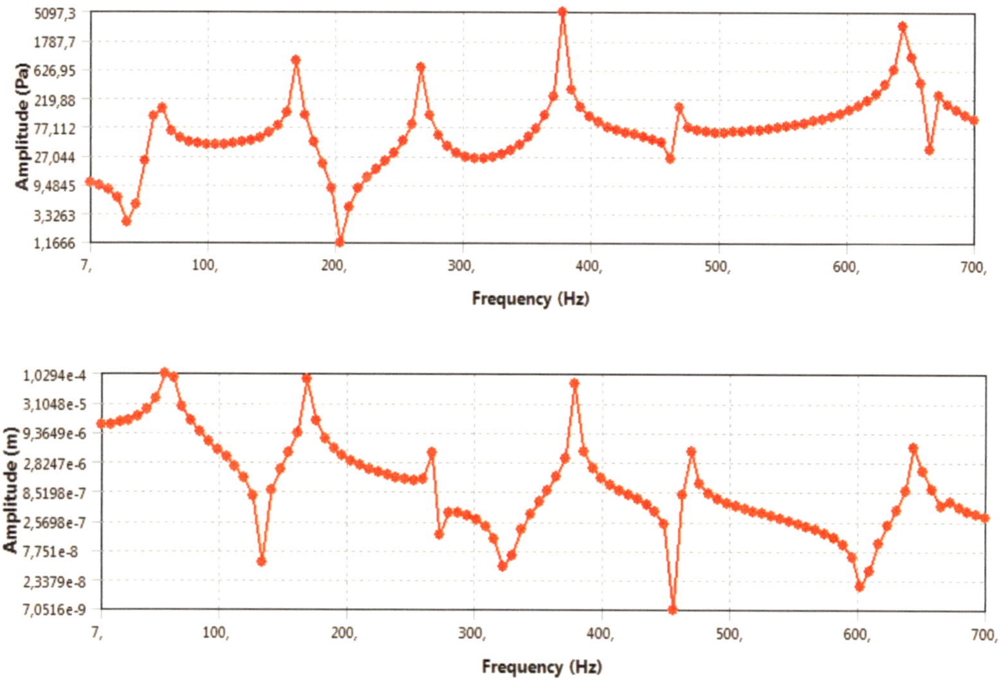

Figure 13. Variation of stress amplitude and displacement amplitude with different exciting frequencies.

specific for all the examined blades. The maximum displacement value indicated by FRF was very small as compared to the maximum displacement value calculated using static analyses of turbine (2.1589e-3 m), under which the turbine blade is safe for operation. In addition, FRF indicated no chances of the occurrence of resonance within this range.

4. Conclusion

The blades represent a vital component of any hydrokinetic turbine due to their complexity, cost, and significant effect on the operating efficiency. During its lifetime, a hydrokinetic turbine blade is subject to different types of loads, such as hydrodynamic, inertial, and gravitational forces. The hydrodynamic design provided the blade external shape, i.e., the chord and twist angle distributions along the blade, which resulted in optimal performance of the hydrokinetic turbine over its lifetime.

The structural design of the blades of a hydrokinetic turbine is paramount to ensure they are rugged, robust, and reliable and yet effective at capturing maximum available energy in the hostile environment of water currents. Therefore, a coupled fluid-structure interaction analysis for a horizontal axis hydrokinetic turbine of 1 kW was carried out using the software ANSYs Workbench to know the effect of several blade structure shapes and different materials on the stresses developed on blades due to the influence of the operating loads. ANSYs-ACP was used as pre-processor for structural analysis of layered composite geometries B and C. FEA indicated that the turbine blade geometries might satisfactorily support the force exerted on it under the combined effects of the centrifugal, gravitational, and hydrodynamic studied loads, since for the analyzed external loads on the blade the stresses were observed to be still below the yield stress of the applied materials, preventing the blades from failure.

Additionally, the dynamic characteristics of the hydrokinetic turbine blade are important to ensure that the blade is rotating at frequencies as far as possible from its natural frequencies to prevent resonance phenomena. Mode shapes generated from modal analysis showed that different frequencies produced different characteristics, including torsional, bending, and combination of torsional and bending modes. From modal analysis and harmonic analysis, it can be observed that the first natural frequencies were far higher than the maximum operating frequencies, leading to a clear indication that the turbine is safe from resonance phenomenon. The harmonic responses within the specified range were also acceptable, as the maximum values of displacement were far lesser than the static displacement values of the analyzed models.

Acknowledgements

The authors gratefully acknowledge the financial support from the Colombian Institute of Science and Technology (COLCIENCIAS) and the Universidad de Antioquia.

Author details

Edwin Chica[1*] and Ainhoa Rubio-Clemente[2]

*Address all correspondence to: edwin.chica@udea.edu.co

1 Departamento de Ingeniería Mecánica, Facultad de Ingeniería, Universidad de Antioquia, Medellín, Colombia

2 Grupo de Diagnóstico y Control de la Contaminación (GDCON), Facultad de Ingeniería, Sede de Investigaciones Universitarias (SIU), Universidad de Antioquia UdeA, Medellín, Colombia

References

[1] M. Anyi and B. Kirke. Evaluation of small axial ow hydrokinetic turbines for remote communities. Energy for Sustainable Development. 2010;**14**(2):110-116.

[2] R.H. van Els and A.C.P.B. Junior. The Brazilian experience with hydrokinetic turbines. Energy Procedia Clean, Efficient and Affordable Energy for a Sustainable Future: The 7th International Conference on Applied Energy (ICAE2015), Abu Dhabi, United Arab Emirates. 2015;**75**:259-264.

[3] E. Rosenberg, A. Lind, and K.A. Espegren. The impact of future energy demand on renewable energy production case of Norway. Energy. 2013;**61**:419-431.

[4] H.J. Vermaak, K. Kusakana, and S.P. Koko. Status of micro-hydrokinetic river technology in rural applications: a review of literature. Renewable and Sustainable Energy Reviews. 2014;**29**:625-633.

[5] M.S. Gney and K. Kaygusuz. Hydrokinetic energy conversion systems: a technology status review. Renewable and Sustainable Energy Reviews. 2010;**14**(9):2996-3004.

[6] M.J. Khan, G. Bhuyan, M.T. Iqbal, and J.E. Quaicoe. Hydrokinetic energy conversion systems and assessment of horizontal and vertical axis turbines for river and tidal applications: a technology status. Applied Energy. 2009;**86**(10):1823-1835.

[7] D. Kumar and S. Sarkar. A review on the technology, performance, design optimization, reliability, techno-economics and environmental impacts of hydrokinetic energy conversion systems. Renewable and Sustainable Energy Reviews. 2016;**58**:796-813.

[8] M. Anyi and B. Kirke. Hydrokinetic turbine blades: design and local construction techniques for remote communities. Energy for Sustainable Development. 2011;**15**(3):223-230.

[9] J.F. Manwell, J.G. McGowan, and A.L. Rogers. Wind Energy Explained: Theory, Design and Application. Wiley, Chichester. 2009.

[10] A.H. Munoz, L.E. Chiang, and E.A. De la Jara. A design tool and fabrication guidelines for small low cost horizontal axis hydrokinetic turbines. Energy for Sustainable Development. Wind Power Special Issue. 2014;**22**:21-33.

[11] J.G. Slootweg, H. Polinder, and W.L. Kling. Dynamic modelling of a wind turbinewith doubly fed induction generator. In Power Engineering Society Summer Meeting, Vancouver, Canada. 2001;**1**:644-649.

[12] G. Riegler. Principles of energy extraction from a free stream by means of wind turbines. Wind Engineering. 1983;**7**(2):115-126.

[13] E. Chica, F. Pérez, A. Rubio-Clemente, and S. Agudelo. Design of a Hydrokinetic Turbine. WIT Transactions on Ecology and the Environment. Wessex Institute of Technology, UK. 2015;**195**:137-148.

[14] E. Chica, F. Pérez, and A. Rubio-Clemente. Rotor structural design of a hydrokinetic turbine. International Journal of Applied Engineering Research. 2016;**11**(4):2890-2897.

[15] J.N. Goundar and M.R. Ahmed. Numerical and experimental studies on hydrofoils for marine current turbines. Renewable Energy 2012;**42**:173-179.

[16] Matweb. Material Property Data. http://www.matweb.com. 2015.

[17] D.M. Grogan, S.B. Leen, C.R. Kennedy, and C.M. Brdaigh. Design of composite tidal turbine blades. Renewable Energy 2013;**57**:151-162.

[18] S.H. Pierson. Composite Rotor Design for a Hydrokinetic Turbine. University of Tennessee Honors Thesis Projects, University of Tennessee Knoxville, Tennessee, 2009.

[19] H. Li, Z. Hu, K. Chandrashekhara, X. Du, and R. Mishra. Reliability-based fatigue life investigation for a medium-scale composite hydrokinetic turbine blade. Ocean Engineering. 2014;**89**:230-242.

[20] S. Saad, A. Behzad, and M. Amir. Dynamic analysis of a 5 KW wind turbine blade with experimental validation. Journal of Space Technology. 2014;**4**(1):82-87.

[21] N. Tenguria, N.D. Mittal, and S. Ahmed. Modal analysis for blade of horizontal axis wind turbine. Asian Journal of Scientific Research 2011; **4**:326-334.

[22] A. Gangele and S. Ahmed. Modal analysis of S809 wind turbine blade considering different geometrical and material parameters. Journal of the Institution of Engineers (India): Series C. 2013;**94**(3):225-228.

[23] C. Liu, D. Jiang, and J. Chen. Vibration characteristics on a wind turbine rotor using modal and harmonic analysis of FEM. In 2010 World Non-Grid-Connected Wind Power and Energy Conference. IEEE, Nanjing, China. 2010:1-5.

Chapter 4

Planning Hydropower Production of Small Reservoirs Under Resources and System Knowledge Uncertainty

Divas Karimanzira, Thomas Rauschenbach,
Torsten Pfuetzenreuter, Jing Qin and Zhao Yun

Additional information is available at the end of the chapter

http://dx.doi.org/10.5772/66912

Abstract

Available energy from water varies widely from season to season, depending on precipitation and streamflows, especially in small catchments. In addition, the reservoir operation problem is associated with the inability of operators to formulate crisp boundary conditions, due to uncertainty in knowledge. In this chapter, an approach for planning the operation of small multipurpose reservoir systems for hydropower generation and flood control under consideration of the stochastic nature of inflows and initial storage levels and allowed formulation of constraints with some range of uncertainty will be presented. The approach is based on joint chance constrained and fuzzy programming, which addresses the problem of including risk directly in the optimization. Therefore, the stochastic nature of inputs is incorporated directly in the model through the use of convolution of random variables. Furthermore, probabilistic/vague constraints and preassigned tolerance levels are used to transform the stochastic optimization problem into its deterministic equivalent. The approach searches for a control strategy, which maximizes the benefits acquired from hydropower generation and minimizes the economic losses incurred due to not meeting the required reliability levels from the various purposes served by the reservoir system. Besides the optimal reservoir release strategy, this approach also determines the optimal reliabilities of satisfying hydropower demand and flood control storage requirements. Therefore, this tool has some advantages in planning the operations of reservoirs in extreme hydrological events such as floods and droughts. The system is applied to the Wuyang small hydropower plants cascade in the People's Republic of China.

Keywords: stochastic optimization, fuzzy programming, uncertainty, hydropower

1. Problem description

Operating small hydropower plant reservoirs is a very difficult task. The managers should make future plans of releasing the water in the reservoir in order to achieve all stakeholders' requirements under consideration of the water availability. For scheduling reservoirs using optimization methods, information of the water coming from the catchment should be forecasted, the initial reservoir water levels are predefined, and the decisions are made on the amount of water to be released. It is prevailing to use past information to deterministically forecast the future, but this is quite erratic due the variability of climate and runoff. Supplementary to this, reservoir managers as humans introduce uncertainty in the interpretation of constraints into the reservoir operation system. This calls upon the consideration of robustness into the optimization system. Constraints can be generally classified into two categories: (1) physical limits and (2) operating limits. A schedule violating physical limit or constraint would not be acceptable. However, operating limits are often introduced to enhance system security, but do not represent physical bounds. Such operating limits can be temporarily violated to a certain extend if necessary, and therefore, they are fuzzy in nature, and crisp treatment of them may lead to over conservative solutions. Crisp constraints are required for the implementation of traditional deterministic optimization models. Therefore, the goal of this work is to take into account the hydrologic variability and allow formulation of constraints with some range of uncertainty.

2. Introduction

Linear programming has been used to solve many real-world problems. This method assumes that the data are definitely known, the constraints are crisp, and the objectives are well defined and can be easily formalized. However, this is not realistic in many situations. Imprecise and vague data make solving many optimization problems difficult. There are different types of uncertainty: (1) uncertainty caused by scarcity of information or (2) that the future state of the system under consideration might not be completely known. This type of uncertainty has been handled by probability theory [1–3].

Archibald et al. [1] use inflow scenarios instead of inflow probability distributions to solve the stochastic optimization problem. Faber and Stedinger [3] also apply this method as well as Schwanenberg et al. [4] who applied the approach in a real-time reservoir operation setup. With streamflow probability distributions, it may take as many state variables to represent the streamflows as there are reservoirs in the cascade. This is the case when there is little or no correlation among concurrent reservoir inflows. This type of problem is difficult to solve with SDP in an amount of time which is reasonable, if several reservoirs reside on the river. It has been shown that the number of state variables in SDP problems can be reduced by applying transformations such as principal component analysis [5–7].

http://dx.doi.org/10.5772/66912

Stochastic linear programming (SLP), in which the inflows are represented by first-order Markov chains, has been developed for optimizing operating strategy of a reservoir [8]. Theoretically, SLP presented in Ref. [8] can be extended to any cascade of reservoirs. However, in practice, the number of reservoirs in the cascade should be small, due to the fact that the computation time increases exponentially with the number of projects in series, as in the case of SDP. Birge [9] showed that a large stochastic multistage linear programming problem can be decomposed into one-stage linear programming problems by applying the Benders' decomposition. Pereira and Pinto [10] used the same method to determine monthly operating policy over half a year for a hydropower system of 37 reservoirs in Brazil. They represented streamflows by scenario trees with two branches in month one, four in month two, and so on. Their method is known as stochastic dual dynamic programming (SDDP).

Linear programming (LP) has been applied to solve implicit stochastic optimization problems. In this case, implicit states that a deterministic problem is solved several times, each time with a different streamflow scenario. To obtain a closed-loop solution, the results achieved from the optimizations are fed to a regression model. This method was also applied by Karamouz et al. [11] to determine a reservoir operating strategy. However, Seifi and Hipel [12] showed that there is no guarantee that the strategy determined will be feasible and efficient enough.

Chance constrained programming (CP) is often applied in reservoir management to reduce the risk of violating the boundary conditions, for example, water level and discharge. But it was shown by Loucks and Dorfman [13] that CP models are very conservative and generate operating policies that exceed the desired reliability levels. However, to solve this problem, Simonovic and Marino [14] developed in their paper a two-step method to determine the best reliability levels. The reliability levels are set in step 1, while the optimal open-loop strategy for operation of the reservoirs is determined in step 2 with linear programming (LP). However, Strycharczyk and Stedinger [15] mentioned in their paper one of the problems with this method that the reservoir discharge in period t is constant, although the problem is stochastic. In stochastic reservoir management, the content of the reservoir is a random variable since it is fed by a streamflow, which is random. The content of the reservoir at start of period t can take any value between the dead water level and the maximum reservoir volume. The optimal reservoir release for a certain period is a function of the reservoir content. Therefore, the open-loop operating strategy described in Ref. [14] not quite acceptable for listing as a stochastic reservoir management problem.

However, CP is applicable to stochastic reservoir management if there is "enough" observed data, so that probability distribution function can be formulated. In some cases, information is deposited in form of expert knowledge. This requires the introduction of the fuzzy model. Bellman and Zadeh [16] introduced the notion of a fuzzy decision making. Recently, a large number of papers have been devoted to the application of fuzzy linear programming (FLP) in modeling and solving problems of real life. Further, Zimmermann [17] and Tanaka et al. [18] applied fuzzy optimization to LP problems with multiple conflicting objectives. Zhang et al. [19] formulated a FLP problem as a four-term objective constrained optimization problem, where the cost coefficients were not crisp.

3. Methodology

3.1. Dealing with uncertainty

Hydrologic processes are random, and thus, the uncertainty as a result of variability may be appropriately quantified using the probabilistic approach. Unfortunately, this approach may lead to unreliable results due to its sole dependency on amount of the available data, the choice of the applied PDFs, inability to deal with knowledge, and human bias. Hence, other methods should be applied in these cases, where the probabilistic approach is partly applicable. The fuzzy method has proven to be very applicable to map systems, which are uncertain and/or include vague expert knowledge. According to the previous information, it is clear that the two methods can produce promising results if they are applied in conjunction. In this chapter, the power of these two approaches is integrated together as illustrated in **Figure 1**.

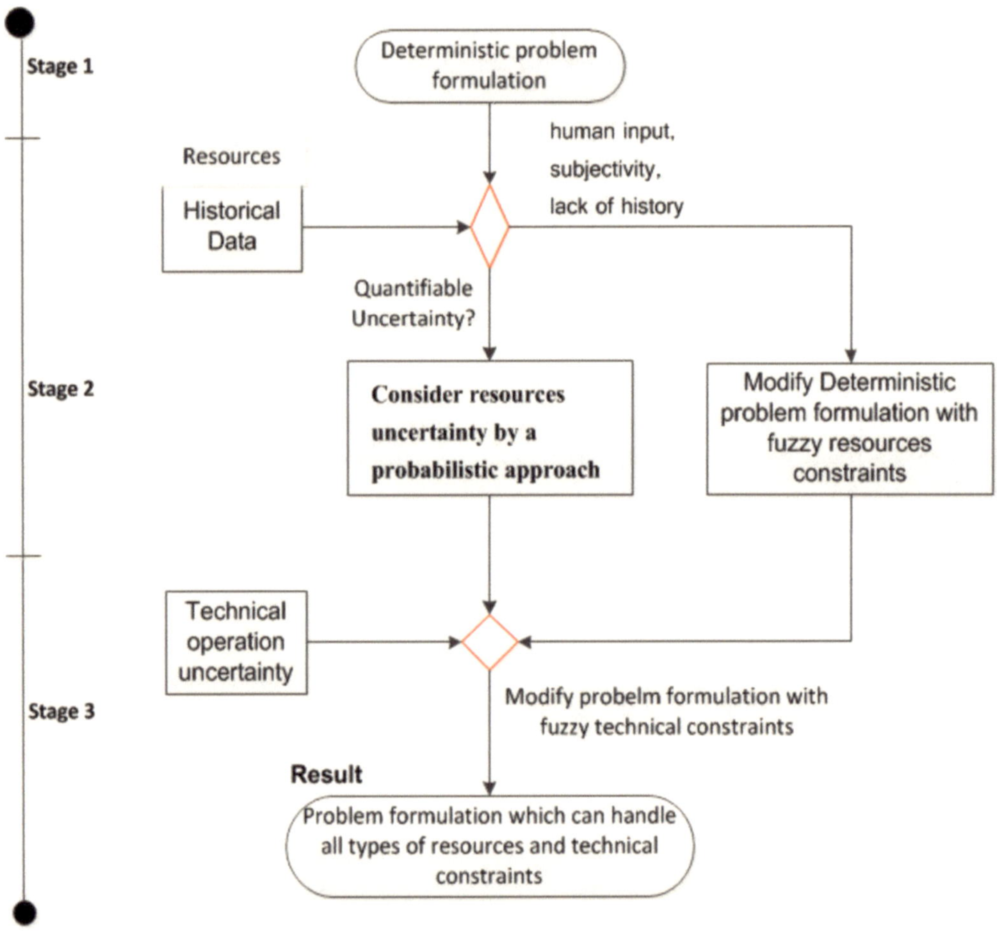

Figure 1. Approach to dealing with different types of uncertainty.

In the first stage, a deterministic optimization of the reservoir cascade is formulated to understand the system. The deterministic formulation is extended in two ways in the second stage depending on the data availability to consider random variables, for example, variability in the inflows and demand. An extension with chance constraints is applied if data are available as option one, an approach that has been extensively used in water resources [20]. In case of historical data scarcity, option 2 applies, whereby the variable resources are considered fuzzy. In addition, in the third stage, the problem will be addressed using a fuzzy optimization approach to include vagueness in the constraints.

3.2. Mathematical formulation

3.2.1. Stage one: understanding the system using a deterministic approach

In **Figure 2**, a flux diagram of a cascade composed of several reservoirs is shown. The deterministic modeling technique enables us to describe all water fluxes as shown in **Figure 3** during every simulation stage.

Note that for each month the storage is calculated for each reservoir taking the difference of total inflows and total outflows. Total outflow is equal to the summation of discharged and spilled water, which is the release of the dam and will flow through to a downstream dam. Total inflow is equal to the summation of the released flow from an upstream reservoir and intermediate flows.

The storage equation is defined in a loop, where the storage at the end of time step "k" is dependent on the storage at the end of time step "k-1_L" the inflow during time step "k_L" and the turbine and spilled flow during "k_L" This is written in the following format:

$$S_i^k = S_i^{k-1} + I_i^k + Q_{i-1}^{k-1} + SP_{i-1}^{k-1} - Q_i^k - SP_i^k \quad (1)$$

for $k = 1, \ldots, N$ and $i = 1, \ldots, M$, S_i^0 is given, $Q_i^0 = 0$ and $SP_i^0 = 0$

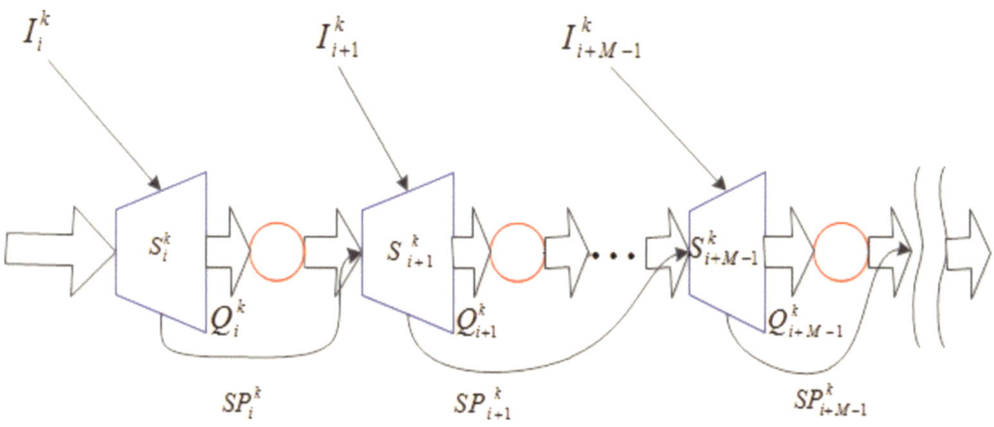

Figure 2. Representation of a cascade with M reservoirs.

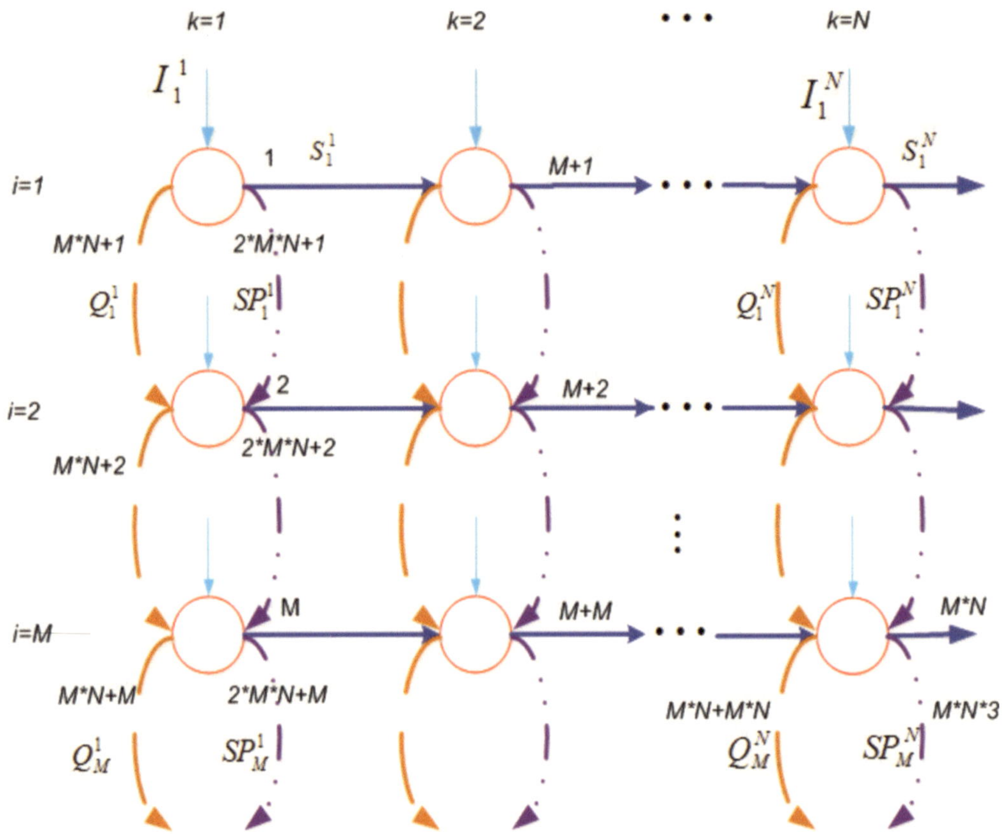

Figure 3. Water fluxes in a N time period.

where S_i^k is reservoir storage volume of reservoir i at the end of period k, I_i^k is intermediate flow (inflow to the reservoir i during period k, apart from the release from an upstream reservoir). Accordingly, it can be seen in **Figure 3** that every stage is composed of three decision nodes. Therefore, to completely determine a given stage, a set of M equations is sufficient:

$$\begin{gathered}
S_1^1 = S_1^0 + I_1^1 - Q_1^1 - SP_1^1 \\
S_2^1 = S_2^0 + I_2^1 + Q_1^1 + SP_1^1 - Q_2^1 - SP_2^1 \\
S_3^1 = S_3^0 + I_3^1 + Q_2^1 + SP_2^1 - Q_3^1 - SP_3^1 \\
\cdots \\
S_1^N = S_1^{N-1} + I_1^N - Q_1^N - SP_1^N \\
S_2^N = S_2^{N-1} + I_2^N + Q_1^{N-1} + SP_1^{N-1} - Q_2^N - SP_2^N \\
S_3^N = S_3^{N-1} + I_3^N + Q_2^{N-1} + SP_2^{N-1} - Q_3^N - SP_3^N
\end{gathered} \tag{2}$$

Parameters determined in these equations must be within real system design ranges. This means that each variable should respect the following constraints:

$$\begin{array}{l} \underline{S}_i \leq S_i^k \leq \overline{S}_i \\ \underline{Q}_i \leq Q_i^k \leq \overline{Q}_i \\ SP_i^k \geq 0 \\ S_i^0 \leq S_i^N \end{array}, \; k = 1, \ldots, N \text{ and } i = 1, \ldots, M \tag{3}$$

where $\underline{S}_i$ and $\overline{S}_i$ are lower and upper bounds of stored water of reservoir i, $\underline{Q}_i$ and $\overline{Q}_i$ are discharged water lower and upper bounds of reservoir i. Even though the volume is not forced to follow a certain reference, the last constraint ensures that the final volume of the reservoir is not below the initial value [21].

Equation systems (1) to (2) are sufficient to linearly describe the operational strategy of any cascade system with M projects. However, this problem can be simplified further by considering that every branch is represented by one state variable. This suggests a transformation of variable that still represents water fluxes dynamics. The next set of Eqs. (4)–(6) shows how to attain this for the first stage for a three-reservoir system, for example:

$$z(1) = S_1^0 + I_1^1 - z(M{*}N + 1) - z(2{*}M{*}N + 1) \tag{4}$$

$$\begin{array}{c} z(2) = S_2^0 + I_2^1 + z(M * N + 1) + z(2 * M * N + 1) \\ -z(M * N + 2) - z(2 * M * N + 2) \end{array} \tag{5}$$

$$\begin{array}{c} z(3) = S_3^0 + I_3^1 + z(M * N + 2) + z(2 * M * N + 2) \\ -z(M * N + 3) - z(2 * M * N + 3) \end{array} \tag{6}$$

Eqs. (7), (8), and (9) are the equivalent for all remaining decision points:

$$z(j) = z(j - i) + I_i^k - z(M * N + j) - z(2 * M * N + j) \tag{7}$$

$$\begin{array}{c} z(j+1) = z(j+1-i) + I_2^k + z(M * N + j - 1) + z(2 * M * N + j - 1) \\ -z(M * N + j) - z(2 * M * N + j) \end{array} \tag{8}$$

$$\begin{array}{c} z(j+2) = z(j+2-i) + I_3^k + z(M * N + j - 1) + z(2 * M * N + j - 1) \\ -z(M * N + j) - z(2 * M * N + j). \end{array} \tag{9}$$

The initial volume (S_i^0) in each reservoir i and its inflows (I_i^k) are known, and together they build the initial data set for this problem. In order to apply these data, in an iterative optimization algorithm, they are represented in vector form as follows:

$$B = \begin{bmatrix} S_1^0 + I_1^1 \\ S_2^0 + I_2^1 \\ S_3^0 + I_3^1 \\ \vdots \\ I_1^N \\ I_2^N \\ I_3^N \end{bmatrix} \quad (10)$$

Consequently, equations are represented in matrices form; therefore, A represents equality constraints for state variables, and finally, Z is the vector of variables:

$$A = \begin{bmatrix} 1 & 0 & 0 & & 0 & 0 & 0 \\ 0 & 1 & 0 & \cdots & 0 & 0 & 0 \\ 0 & 0 & 1 & & 0 & 0 & 0 \\ & \vdots & & \ddots & & \vdots & \\ 0 & 0 & 0 & & 1 & 0 & 0 \\ 0 & 0 & 0 & \cdots & -1 & 1 & 0 \\ 0 & 0 & 0 & & 0 & -1 & 1 \end{bmatrix}; \; Z = \begin{matrix} z(1) \\ z(2) \\ z(3) \\ \vdots \\ z(N_m - 2) \\ z(N_m - 1) \\ z(N_m) \end{matrix} \quad (11)$$

The problem formulation is completed by designing the optimization criteria (objective function). The objective function for hydropower energy maximization can be expressed as a product of the head for hydropower generation and the release. Therefore, nonlinear programming (NLP) may be considered for solving this problem. However, this research takes a different approach by applying linear programming to the linear operation model.

Theoretically, hydropower capacity of a storage plant installed at a reservoir can be expressed as

$$P = \eta \rho g Q(t) H_n(t) = \gamma Q(t) H_n(t) \quad (12)$$

where $\eta \rho g$, γ can be a constant value, η is the overall, ρ is the water density, and g is the gravity acceleration; $Q(t)$ is the release for the power generation of at time tL and the net head $H_n(t)$ can then be written as

$$H_{n(t)} = H(t) - H_{tail}(t) - H_{loss}(t) \quad (13)$$

where $H(t)$ denotes reservoir water level, $H_{tail}(t)$ is the water level of the downstream, and $H_{loss}(t)$ denotes the head loss at a given instant t. If $H_{tail}(t)$ and $H_{loss}(t)$ are negligible in comparison with $H(t)$, the approximation $H_n(t) \cong H(t)$ is applied, and the objective function for maximizing hydropower energy can be formulated as

http://dx.doi.org/10.5772/66912

$$E = \gamma \sum_{k=1}^{N} Q(k)H(k) = \gamma\, \mathbf{Q} \cdot \mathbf{H} \tag{14}$$

where k is the discrete time and $Q(k)$ and $H(k)$ are the releases for hydropower generation and reservoir water level at time k. N is the total time. Eq. (14) is the nonlinear product of the vectors **Q** and **H**. However, in this paper, a linear function is used based on the fact that Eq. (15) is maximized by maximizing each member of $\mathbf{Q}[Q(1), Q(2), \ldots, Q(J)]$ and $\mathbf{H}[H(1), H(2), \ldots, H(N)]$.

$$\mathbf{Q} \cdot \mathbf{H} = Q(1)H(1) + Q(2)H(2) + \cdots + Q(N)H(N) \tag{15}$$

In turn, the nonlinear Eq. (14) can be replaced by the linear Eq. (16) as an objective function Z for maximizing hydropower energy. Eq. (16) linearly sums up each element of **Q** and **H**.

$$\max_{Q,H} w_H \sum_{k=1}^{N} H(k) + w_Q \sum_{k=1}^{N} Q(k) \tag{16}$$

where w_H and w_Q are penalty factors on water level and the releases, respectively. Eq. (17) can replace Eq. (16) due to the fact that $H(j)$ is directly proportional to $S(j)$, that is, w_H can be substituted by w_S as:

$$\max_{Q,S} \sum_{i=1}^{M} \left(w_S \sum_{k=1}^{N} Q(k) + w_Q \sum_{k=1}^{N} S(k) \right) \tag{17}$$

Eq. (17) is the linear combination of reservoir storage $S(j)$ and release $Q(j)$, and it can be represented as an alternative objective function for maximizing hydropower energy.

Subject to

$$\begin{gathered} A \cdot Z = B \\ \underline{S}_i \le S_i^k \le \overline{S}_i \\ \underline{Q}_i \le Q_i^k \le \overline{Q}_i \\ SP_i^k \ge 0 \\ S_i^0 \le S_i^N \end{gathered} \tag{18}$$

The constraints are linear, the state equation is linear, and the objective function is chosen in linear form. The optimal solution can be obtained using various software tools readily available. Time lag or routing of flows between reservoirs is neglected in this formulation, which is reasonable for monthly time steps. Rainfall and net losses due to seepage, evaporation, and other reservoir losses are subtracted directly from the river inflows.

After formulating the problem deterministically, stage 2 follows which considers the uncertainty in the available resources. As shown in **Figure 2**, stage 2 is performed in two options according to the quantifiability of the available data.

3.2.2. Consideration of resources uncertainty by a probabilistic approach

To deal with LP under uncertainty, the chance constrained programming was introduced by Charnes and Cooper [22]. It extends the LP to enable the violation of the constraints to a certain extend. The reliability $\alpha \in [0, 1]$ of not violating a constraint is specified by the decision maker, and thus, it allows for decision maker to directly control the level of risk he/she finds acceptable.

The deterministic reservoir model developed previously will be transformed here in the probabilistic form to deal with some uncertain inputs. The transformation to stochastic optimization is done through the introduction of an additional probabilistic constraint, which is shown below

$$P\{\tilde{S}_i^k \le S_{i,target}\} \ge \propto,\ k = 1, \ldots, N \text{ and } i = 1, \ldots, M \tag{19}$$

where $\tilde{S}_i k$ is the random equivalent of S_i^k, the storage at the end of period k, $S_{i,\ target}$ is the known decision maker specified target storage level of the reservoir, and α is the decision maker specified reliability of not violating constraint Eq. (19). It takes values between 0 and 1. This formulation is the realistic representation of the cascade control problem since in practice, bounds on storage are often violated by expansion of the conservation pool into the flood control pool. Adopting chance constraints simplifies the stochastic problem to deterministic. The random variables $\tilde{I}_i^k$ and $\tilde{S}_i 0$ are additive in consecutive periods, which makes the application of the convolution method feasible, whether they are linearly correlated or not [23]. The following procedure is used to do the transformation of the stochastic constraint.

Step 1. Eq. (1) is substituted into Eq. (19) to form:

$$P\{\tilde{S}_i^k = S_i^{k-1} + \tilde{I}_i k + Q_{i-1}^k + SP_{i-1}^k - Q_i^k - SP_i^k \le S_{i,target}\} \ge \propto \tag{20}$$

where $\tilde{I}_i^k$ is the random equivalent of I_i^k, the inflow during $k = 1, \ldots, N$ and $i = 1, \ldots, M$.

Step 2. A deterministic equivalent of the Eq. (20) is found by inversion and rearrangement leading to:

$$S_{i,target} - S_i^{k-1} + \tilde{I}_i k + Q_{i-1}^k + SP_{i-1}^k - Q_i^k - SP_i^k \le F_{x_i^k}^{-1}(1 - \propto) \tag{21}$$

where for $k = 1, \ldots, N$ and $i = 1, \ldots, M$, $F_{x_i^k}^{-1}(1 - \propto)$ is the inverse value of the cumulative distribution function of the convoluted $\tilde{I}_i^k$, evaluated at (1-α). Henceforth, it will be replaced by $x_i^{k,1-\alpha}$.

Step 3. The expression for two deterministic chance constraint time steps is given below,

for $k = 1$ and $i = 1, \ldots, M$,

$$
\begin{aligned}
&P\left\{\tilde{S}_i^1 \leq S_{i,target}\right\} \geq \propto \\
&P\left\{\tilde{S}_i^0 + \tilde{I}_i 1 + Q_{i-1}^1 + SP_{i-1}^1 - Q_i^1 - SP_i^1 \leq S_{i,target}\right\} \geq \propto \\
&P\left\{\tilde{I}_i^1 \leq Q_i^1 + SP_i^1 + S_{i,target} - \tilde{S}_i 0 - Q_{i-1}^1 - SP_{i-1}^1\right\} \geq \propto \\
&\quad x_i^{1,1-\alpha} \geq Q_i^1 + SP_i^1 + S_{i,target} - \tilde{S}_i^0 - Q_{i-1}^1 - SP_{i-1}^1
\end{aligned} \tag{22}
$$

for $k = 2$ and $i = 1, \ldots, M$,

$$
\begin{aligned}
&P\left\{\tilde{S}_i^2 \leq S_{i,target}\right\} \geq \propto \\
&P\{S_i^1 + \tilde{I}_i^2 + Q_{i-1}^2 + SP_{i-1}^2 - Q_i^2 - SP_i^2 \leq S_{i,target}\} \geq \propto
\end{aligned} \tag{23}
$$

Substituting for S_i^1

$$
\begin{aligned}
&P\left\{\begin{array}{c} (\tilde{S}_i^0 + \tilde{I}_i^1 + Q_{i-1}^1 + SP_{i-1}^1 - Q_i^1 - SP_i^1) \\ +\tilde{I}_i^2 + Q_{i-1}^2 + SP_{i-1}^2 - Q_i^2 - SP_i^2 \leq S_{i,target} \end{array}\right\} \geq \propto \\
&P\left\{\begin{array}{c} \tilde{I}_i^1 + \tilde{I}_i^2 \leq Q_i^1 + SP_i^1 - Q_{i-1}^1 - SP_{i-1}^1 \\ -Q_{i-1}^2 - SP_{i-1}^2 + Q_i^2 + SP_i^2 + S_{i,target} - \tilde{S}_i^0 \end{array}\right\} \geq \propto \\
&x_i^{2,1-\alpha} \geq Q_i^1 + SP_i^1 - Q_{i-1}^1 - SP_{i-1}^1 - Q_{i-1}^2 \\
&-SP_{i-1}^2 + Q_i^2 + SP_i^2 + S_{i,target} - \tilde{S}_i^0
\end{aligned} \tag{24}
$$

Eq. (21) can thus be expressed in final simplified chance constraint deterministic form as:

$$
S_{i,target} - \tilde{S}_i 0 + \sum_{k=1}^{N}(Q_i^k + SP_i^k) - \sum_{k=1}^{N}(Q_{i-1}^k + SP_{i-1}^k) \leq x_i^{k,1-\alpha} \tag{25}
$$

for $k = 1, \ldots, N$ and $i = 1, \ldots, M$.

It is important to note that the random variables inflow and initial storage are summed here. For the time interval k = 1, the sum is $\tilde{I}_i 1 + \tilde{S}_i 0$, for k = 2 it is $\tilde{S}_i 0 + \tilde{I}_i 1 + \tilde{I}_i 2, \ldots$, and for k = N it is $\tilde{S}_i 0 + \tilde{I}_i 1 + \tilde{I}_i 2 + \cdots + \tilde{I}_N$. In the two-step algorithm developed by Simonovic to determine the best reliability levels for chance constraints, reliability levels are set in the first step, while the optimal open-loop reservoir operating strategy is determined by linear programming in step 2 [14]. This operating strategy sets the discharge from reservoir i in period t equal to x_t^i. The discharges x_t^i (for all i and t) are then used to transform the probabilistic constraints into deterministic ones. Strycharczyk and Stedinger [15] in their paper stressed out the problem with Simonovic and Marino [14] method that the discharge from a given reservoir i in a given period t is a constant, although the problem is stochastic. In stochastic reservoir management, the content of the reservoir is a random variable since it is fed by a streamflow, which is random. The content of the reservoir at start of period t can take any value between the dead water level and the maximum reservoir volume. The optimal reservoir release for a certain period is a function of the reservoir content. Therefore, the open-loop operating strategy described in Ref. [14] is not quite acceptable for listing as a stochastic reservoir management problem. Therefore, in this paper we consider the initial reservoir storage level as a random variable also.

The random variables inflow and initial storage have known marginal probability distribution functions (PDF), $f(I_k)$ and $f(S_{0,\ i})$, respectively, as a result of fitting a distribution to available historical data. However, the distributions of the sums have to be found. This is accomplished through a step-by-step iterative convolution method:

For $k = 1$, iterative convolution is used to reduce the two random variables $\tilde{I}_i^k$ and $\tilde{S}_i 0$ appearing in the equation of state to a single random variable. The new random variable is obtained as the sum of the random inflow and the random initial storage. Since their probability density functions are known, that of the new variable is obtained by convolution, which is expressed in discrete form by

$$P\{S = r\} = \sum_{i-j=r} p_i^I p_j^S \tag{26}$$

where $r_{\min} \le r \le r_{\max}$, p_i^I is the probability distribution of inflow, p_j^S is the probability distribution of the initial storage. The summation is carried out over the variable I, and hence, expression (26) yields:

$$P\{S = r\} = \sum_{i=a}^{b} P_i^I P_{i-r}^S. \tag{27}$$

The magnitudes of r_{min} and r_{max} are found from $r_{min} = i_{min} - j_{max} = a - d$ and $r_{max} = i_{max} - j_{min} = b - c$ under the constraints $a \le i \le b$ and $c \le j \le d$.

From $k = 2$ to $k = N$, the recursive equation for convolution can be expressed in genral form as in [24] as follows:

$$p_r(k) = \sum_{i-j=r} P\{i_k = i\} p_j(k-1) \tag{28}$$

$$r_{\min} \le r \le r_{max} \tag{29}$$

where $r_{min} = i_{min} + j_{min}$, $r_{max} = i_{max} + j_{max}$, under the constraints: $a \le i \le b$ and $c \le j \le d$.

The problem formulation becomes similar to linear formulation in the deterministic approach; Eq. (6) with the addition of the deterministic chance constraint and Eq. (25) can be solved with the same linear programming approach as the deterministic model formulation.

3.2.3. Stage two and three: consideration of resources and technical uncertainty using a fuzzy approach

If in stage 2 historical data are scarce, we can apply a fuzzy approach for both resources and technical uncertainty. The reservoir operation optimization model formulation will be expanded to utilize the fuzzy linear optimization approach, and in doing so, it will depart from the classical assumptions that all coefficients of the constraints need to be crisp numbers [25]. In the present section, a FLP formulation based on the work of [25] and further by Tanaka et al. [18] that considers both technological coefficients and resources characterized by uncertainty is presented in brief as follows:

The fuzzy version of the traditional linear programming optimization problem presented in Eq. (29) is:

$$\begin{aligned} cx &\lesssim z_0 \\ Ax &\lesssim b \\ x &\geq 0 \end{aligned} \tag{30}$$

The manager's targets and system boundaries are the inequalities in the fuzzy system. The equation expressed that the manager's targets can be lower than his/her desired level z_0. The same applies to the boundaries that they should be in the tolerance level b. The importance of targets and the constraints is set at the same level. For fully symmetric objective and constraints, Zimmermann [26] formulated the problem in simplified form as follows:

$$\begin{aligned} Bx &\lesssim d \\ x &\geq 0 \end{aligned} \tag{31}$$

where $B = \begin{bmatrix} c \\ A \end{bmatrix}, d = \begin{bmatrix} z_0 \\ b \end{bmatrix}$.

The following expression for the (monotonically decreasing) linear membership function was proposed by Zimmerman for the jth fuzzy inequality $(Bx)_j \lesssim d_j$.

$$\mu_j\left((Bx)_j\right) = \begin{cases} 1 & \textit{if } (Bx)_j < d_j \\ 1 - \left((Bx)_j - d_j\right)/p_j & \textit{if } d_j \leq (Bx)_j \leq d_j + p_j \\ 0 & \textit{if } (Bx)_j \geq b_j + p_j \end{cases} \tag{32}$$

where d_j and p_j are the desired level and the tolerance for violation of the jth inequality, respectively. A jth membership function value of 1 denotes that the targets and constraints are fully satisfied. Violating the constraint outside the tolerance band, p_j gives a membership value of 0 and in-between values are linear. The membership function of the fuzzy set "decision" of model in Eq. (6) including the linear membership functions is shown below. The problem of finding the maximum decision is to choose x^* such that

$$\mu_D(x^*) = \max_{x \geq 0} \min_{j=0,\ldots,m} \left\{\mu_j\left((Bx)_j\right)\right\}. \tag{33}$$

In other words, the problem is to find the $x^* \geq 0$ which maximizes the minimum membership function value. This value satisfies the fuzzy inequalities, $(Bx)_j \lesssim d_j$ with the degree of x^* [18].

Substituting the expression (31) for linear membership function into Eq. (32) yields

$$\mu_D(x^*) = \max_{x\geq 0}\ \min_{i=0,\ldots,m} \left\{1 + \frac{d_j}{p_j} - \frac{(Bx)_j}{p_j}\right\}. \tag{34}$$

The fuzzy set for decision can be transformed to an equivalent conventional linear programming problem by introducing the auxiliary variable λ:

$$\text{Maximize } \lambda \tag{35}$$

subject to $\lambda \leq 1 + \frac{d_j}{p_j} - \frac{(Bx)_j}{p_j}$

$$x \geq 0$$

It should be emphasized that the above formulation is for a minimization of the objective function and less than constraints and thus should be modified appropriately for other conditions.

Using the fuzzy optimization approach just described, and using the deterministic model given by Eq. (6) with modification for considering linear membership function for "greater than" constraints, the fuzzy formulation becomes:

$$\text{Maximize } \lambda \tag{36}$$

Subject to

$$\frac{(Bx)_j}{p_i} + \leq 1 + \frac{d_j}{p_j}\ \mathrm{i} = 0, \ldots, 36 \tag{37}$$

$$\frac{(Bx)_j}{p_i} - \leq \frac{d_j}{p_j}\ \mathrm{i} = 37, \ldots, 60 \tag{38}$$

$$x \geq 0$$

Expanding by substituting for $(Bx)_j$

$$\frac{\sum_{i=1}^{M}\left(w_S \sum_{k=1}^{N} Q(k) + w_Q \sum_{k=1}^{N} S(k)\right)}{p_j} + \leq 1 + \frac{d_j}{p_j}\ \mathrm{j} = 0 \tag{39}$$

all other constraints for $k = 1, \ldots, N$ and $i = 1, \ldots, M$

$$\frac{S_i^k - S_i^{k-1} + Q_i^k + SP_i^k - Q_{i-1}^k - SP_{i-1}^k}{p_j} + \leq 1 + \frac{d_j}{p_j} \tag{40}$$

$$j = 1, \ldots, M * N$$

http://dx.doi.org/10.5772/66912

$$\frac{S_i^k}{p_j} + \leq 1 + \frac{d_j}{p_j},\ j = M*N+1, \ldots, 2*M*N \tag{41}$$

$$\frac{Q_i^k}{p_j} + \leq 1 + \frac{d_j}{p_j},\ j = 2*M*N+1, \ldots, 3*M*N \tag{42}$$

$$\frac{S_i^k}{p_j} - \geq \frac{d_j}{p_j},\ j = 3*M*N+1, \ldots, 4*M*N \tag{43}$$

$$\frac{S_i^k - S_i^{k-1} + Q_i^k + SP_i^k - Q_{i-1}^k - SP_{i-1}^k}{p_j} - \geq \frac{d_j}{p_j},\ j = 4*M*N+1, \ldots, 5*M*N \tag{44}$$

$$Q_i^k \geq 0 \tag{45}$$

$$SP_i^k \geq 0 \tag{46}$$

4. Numerical example

The following demonstrates the application of the methodology for a reservoir cascade composed of three projects. All stages will be shown, from deterministic to its modification for the implementation of the probabilistic and fuzzy domains.

4.1. Case study

The reservoir cascade optimization case study is the Wuyang cascade system in the People's Republic of China. An optimization problem is formulated for 12-month time period ($k = 12$) as discussed in preceding sections and solved using data provided from the Institute of Hydraulic and Water Research. The available data are listed in **Table 1**. It is from 1953 to 2009 and includes the constraints in storage capacity and the stream flows.

	Reservoir 1	Reservoir 2	Reservoir 3
Maximum reservoir capacity (m^3)	1.373E8	6.572E8	7.1E7
Dead or minimum reservoir storage, S_{min} (m^3)	500000.0	727000.0	792000.0
Sill of dam elevation operator goal storage, S_{target} (m^3)	1.012E8	4.872E8	5.34E7
Initial storage, S_0 (m^3)	5.12E7	1.206E8	4.48E7
Maximum possible release for non-flooding condition, R_{max} (m^3/s)	50	50	50

Table 1. Hydropower projects data.

5. Results

The cumulative distribution values for $f(I_k)$ and $f(S_{0,\ i})$, obtained by the summation of PDFs with reliability tolerance of 0.9, are shown in **Figure 4**.

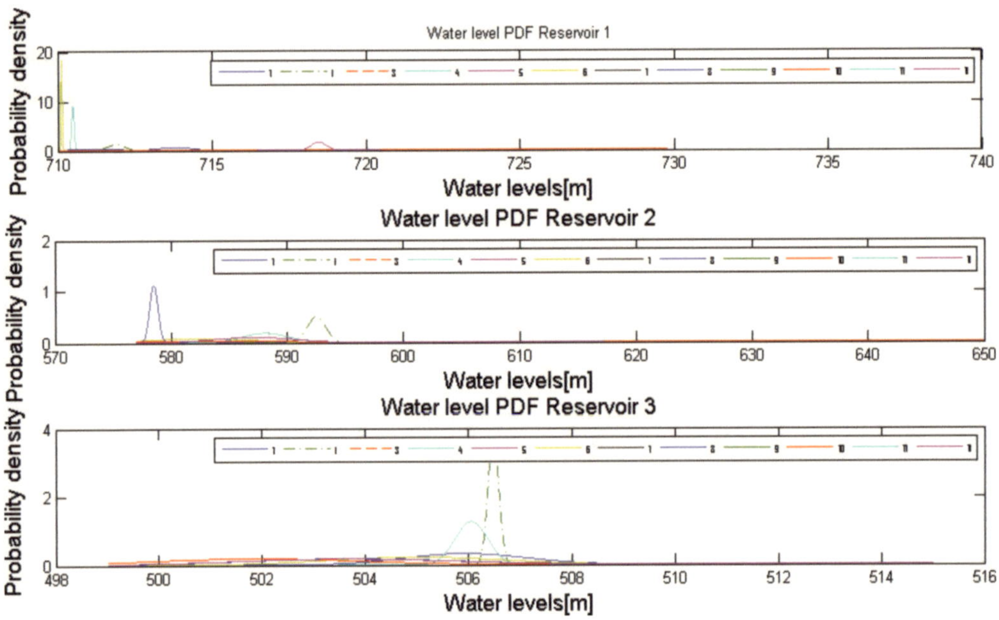

Figure 4. Probability distribution functions of the sum of and .

It is also considered that the water managers wanted some flexibility in the constraint to account for the uncertainty in knowledge, which is not available with the sharp constraint requirements of the deterministic model. Further, the water managers assessed that the annual maximum acceptable storage to mitigate damage due to flood should not exceed S_{flood}. The managers formulated there constraints according to their experience and comfort. Therefore, the FLP approach was chosen to facilitate this insight, and LB and UB of d_j and p_j were determined.

The results in **Figure 5–Figure 7** include a series of release rules for operating period of the 12 months that reservoir operators can follow in order to fulfill the defined objective.

In stage 1, using the deterministic optimization approach and substituting the given data, the above problem with 144 balance equations and 145 constraints becomes readily solvable using linear programming. The optimal solution is shown in **Figure 5**.

In stage 2, the probabilistic optimization approach was applied. Firstly, a PDF was selected from experiments and experience from the structure of the stream flows. Stream flows are positive, and there variance changes proportionally to their mean, which is characteristic of lognormal functions. α was set to 0.9, and the addition of the stream flows and possible initial water levels was determined using a convolution process as described previously in this chapter. Once the convolution process is complete and inflow convoluted values

http://dx.doi.org/10.5772/66912

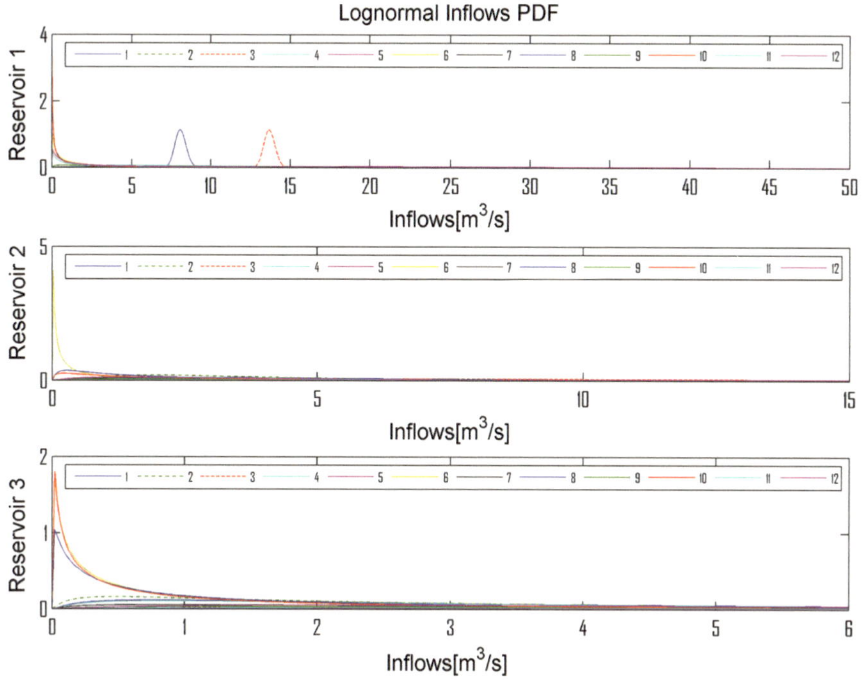

Figure 5. Results of the deterministic method (objective function Z = 0.3857 m^3).

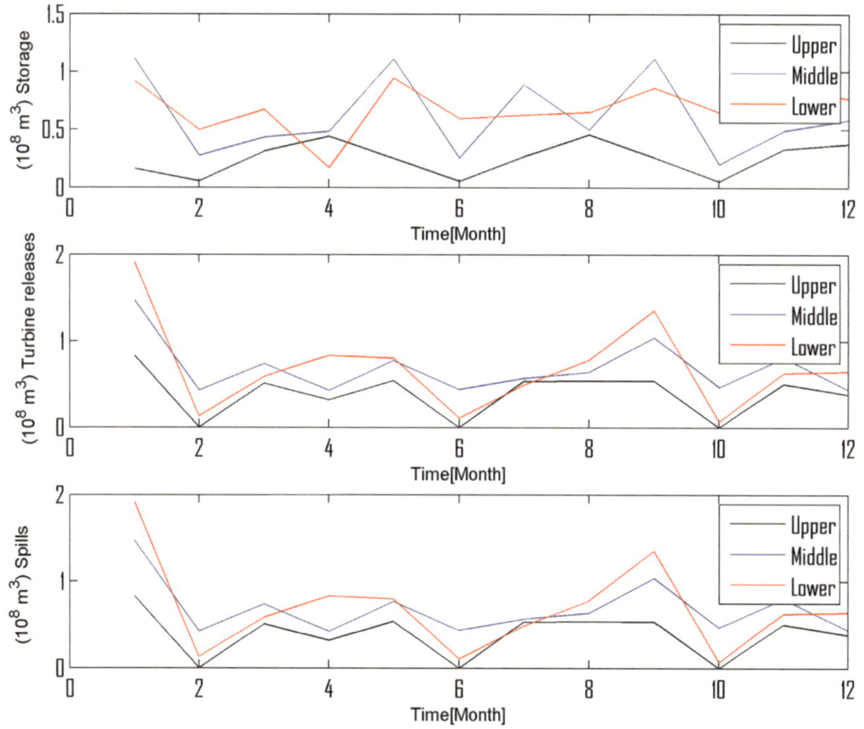

Figure 6. Chance constrained method (objective function Z = 1.4643 m^3).

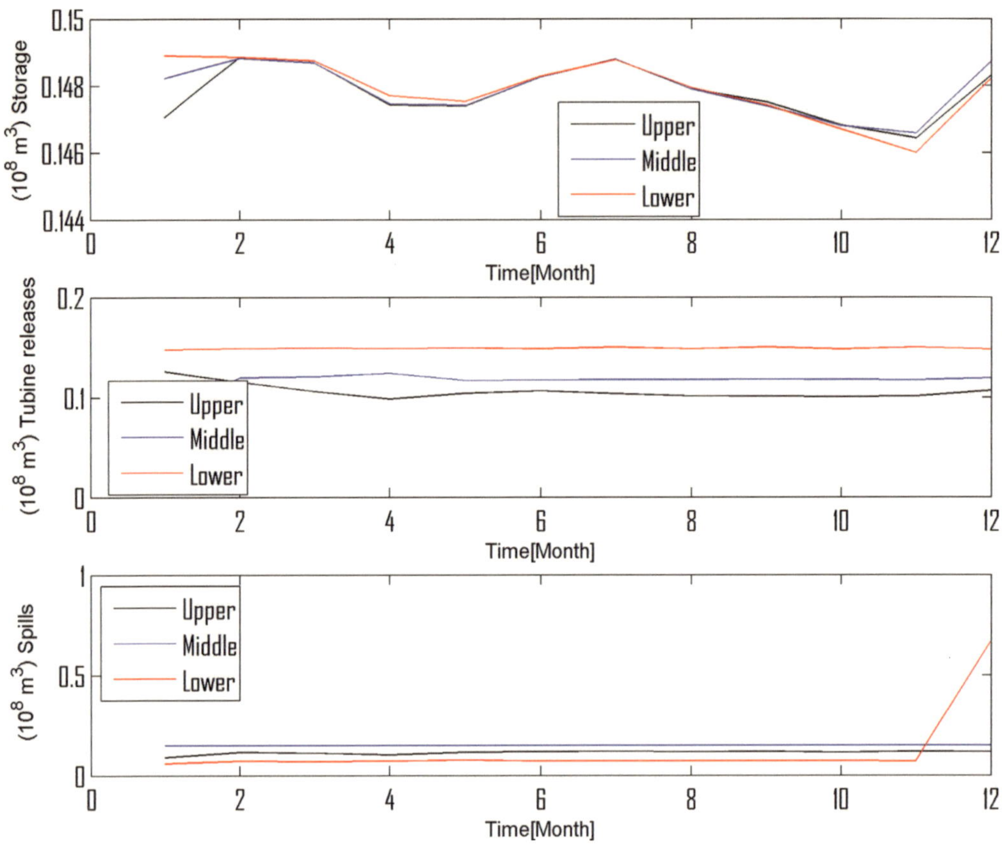

Figure 7. Fuzzy chance constrained method ($Z = 4.4515\ m^3$).

corresponding to the reliability index selected are found, the problem was solved using linear optimization as in the case of the deterministic formulation. The optimal solution is shown in **Figure 6** for reliability level $\alpha = 0.9$.

In the final stage, using fuzzy optimization approach, the values of the tolerance interval d_j and spread of tolerance p_j are substituted into the constraints and the problem is solved using a linear programming solver. The resulting $\lambda = 0.0626$ with corresponding storage and release is shown in **Figure 7**.

6. Discussion

It could be seen that the problem could be solved by all the three methods with different results. However, the linear approach uses water spillage to achieve optimal solution. The chance constraint method offers a way to include reliability in formation of the optimization problem. The requirements of the water managers to have some flexibility in the constraint to account for the uncertainty in knowledge could be realized satisfactory with the fuzzy method.

7. Conclusions

Crisp-defined boundaries do not often resemble real-life situations. It is a fact that uncertainty in decision-making processes occurs at every stage. Therefore, it is necessary to understand uncertainties, which in consequence require understanding of their sources. In water resource management (WRM), uncertainty can be put into two categories: (1) uncertainty due to inherent variability in hydrology and uncertainty contributed by scarcity of data and knowledge. The transformation of deterministic problems into the probabilistic and fuzzy domains was presented in this paper. The fuzzy approach integrates the managers indirectly into the optimization process through their expert information. As shown in this chapter, the probabilistic method is very good in dealing with quantifiable uncertainty. However, its robustness to handle different sources of uncertainties is not sufficient to justify its use under all circumstances. Caution must be taken, pending on the level of precision desired the stochastic or even the deterministic approach may be the better alternative. But the probabilistic approach can be implemented only if uncertainties are quantifiable and sufficient historical data are available. The holic view of integrating the two approaches gives good opportunities to solve problems of reservoir optimization in case of vast data availability, data scarcity, and availability of information from expert knowledge and experience, which cannot be quantified. To show the robustness of the approaches, they were applied to a real case of the Wuyang river cascade in PR China. Here, these data are scarce, and most of the decisions are made subjectively. The obtained results are quite promising. In further sensitivity tests performed (not part of this paper), it could be shown that the combination of the CP and the FLP method is very robust to changes in stream flows, initial reservoir levels, and formulated targets.

Author details

Divas Karimanzira[1]*, Thomas Rauschenbach[1], Torsten Pfuetzenreuter[1], Jing Qin[2] and Zhao Yun[2]

*Address all correspondence to: divas.karimanzira@iosb-ast.fraunhofer.de

1 Department of Surface and Maritime Systems, Fraunhofer I0SB-AST, Ilmenau, Germany

2 China Institute of Water Resources and Hydropower Research (IWHR), Beijing, China

References

[1] Archibald, T. W., McKinnon, K. I. M., and Thomas, L. C. (1997). An aggregate stochastic dynamic programming model of multireservoir systems. Water Resources Research, 33 (2), 333–340. doi:10.1029/96WR02859

[2] Simonovic, C. P., and Peck, A. (2009). Updated rainfall intensity duration frequency curves for the city of London under changing climate. Water Resources Research Report

No. 065, Facility for Intelligent Decision Support, Department of Civil and Environmental Engineering, London, Ontario, Canada, 64 p. ISBN: (print) 978-0-7714-2819-7; (online) 987-0-7714-2820–3.

[3] Faber, B. A., and Stedinger, J. R. (2001). Reservoir optimization using sampling SDP with ensemble streamflow prediction (ESP) forecasts. Journal of Hydrology, 249(1–4), 113–133.

[4] Schwanenberg, D., Xu, M., Ochterbeck, T., Allen, C., and Karimanzira, D. (2014). Short-term management of hydropower assets of the Federal Columbia River power system. Journal of Applied Water Engineering and Research, 2(1), 25–32.

[5] Saad, M., Turgeon, A., Bigras, P., and Duquette, R. (1994). Learning disaggregation techniques for the operation of long-term hydroelectric power systems. Water Resources Research, 30(11), 3203–3295.

[6] Saad, M., Turgeon, A., Bigras, P., and Duquette, R. (1996). Fuzzy learning decomposition for the scheduling of hydroelectric power systems. Water Resources Research, 32(1), 179–186.

[7] Saad, M., and Turgeon, A. (1988). Application of principal component analysis to long-term reservoir management. Water Resources Research, 24(7), 907–912.

[8] Loucks, D. P. (1968). Computer models for reservoir regulations. Journal of the Sanitary Engineering Division. American Society of Civil Engineers, 94(SA4), 657–669.

[9] Birge, J. R. (1985). Decomposition and partitioning methods for multistage stochastic linear programs. Operations Research, 33(5), 989–1007.

[10] Pereira, M. V. F., and Pinto, L. M. V. G. (1985). Stochastic optimization of a multireservoir hydroelectric system: a decomposition approach. Water Resources Research, 21(6), 779–792.

[11] Karamouz, M., Houck, M., and Delleur, J. (1992). Optimization and simulation of multiple reservoir systems. Journal of Water Resources and Planning Management, 118(1), 71–81.

[12] Seifi, A., and Hipel, K. (2001). Interior-point method for reservoir operation with stochastic inflows. Journal of Water Resources and Planning Management, 127(1), 48–57.

[13] Loucks, D. P., and Dorfman, P. (1975). An evaluation of some linear decision rules in chance-constrained models for reservoir planning and operation. Water Resources Research, 11(6), 777–782.

[14] Simonovic, S., and Marino, M. (1982). Reliability programming in reservoir management: 3. Systems of multipurpose reservoirs. Water Resources Research, 18(4), 735–743.

[15] Strycharczyk, J. B., and Stedinger, J. R. (1987). Evaluation of a "reliability programming" reservoir model. Water Resources Research, 23(2), 225–229.

[16] Bellman, R., and Zadeh, L. (1970). Decision making in a fuzzy environment. Management Science, 17, 141–164.

[17] Zimmermann, H. J. (1976). Description and optimization of fuzzy systems. International Journal of General Systems, 2(4), 209–215.

[18] Tanaka, H., Ichihashi, H., and Asai, K. (1991). Formulation of fuzzy linear programming problem based on comparison of fuzzy numbers. Control Cybernetics, 3(3), 185–194.

[19] Zhang, G., Yong-Hong, W. U., Remias, M., and Lu, J. (2003). Formulation of fuzzy linear programming problems as four-objective constrained optimization problems. Applied Mathematics and Computation, 139, 383–399.

[20] Simonović, S. P. (2002). Understanding risk management. In: Proceedings of the Annual Conference of the Canadian Society for Civil Engineering, Montréal, Québec, Canada (pp. 1–12). http://pedago.cegepoutaouais.qc.ca/media/0260309/0378334/SCGC BON/Documents/GE108-Simonovic.pdf.

[21] ReVelle, C. (1999). Optimizing Reservoir Resources: Including a New Model for Reservoir Reliability. New York: Wiley.

[22] Charnes, A., and Cooper, W. (1959). Chance-constrained programming. Management Science, 6(1), 73–79.

[23] Curry, L. G., Helm, J. C., and Clark, R. A. (1973). Chance–constrained model of system of reservoirs. Journal of the Hydraulics Division, 99(12), 2353–2366 (ASCE, 12, HY).

[24] Simonović, S. P. (1979). Two-step algorithm for design-stage long-term control of a multipurpose reservoir. Advanced Water Resources, 2, 47–49 (118).

[25] Zimmermann, H. J. (1996). Fuzzy Set Theory and Its Applications. 3rd edition. Boston, MA: Kluwer Academic Publishers.

[26] Zimmermann, H. J. (1978). Fuzzy programming and linear programming with several objective functions. Fuzzy Sets Systems, 1, 45–55.

Chapter 5

Hydropower Development in Nepal - Climate Change, Impacts and Implications

Ramesh Prasad Bhatt

Additional information is available at the end of the chapter

http://dx.doi.org/10.5772/66253

Abstract

Nepal has endowed high potential of water resources, covering 395,000 ha (48%) area within 45,000 km in length of 6000 rivers with 170 billion m^3 annual runoff and 45,610 MW feasible hydroelectricity generation. Since 1911, 500 kW power generation at Pharping, now reached 782.45 MW production in 2016. Nepal government has planned to increase its current 67.3% access in electricity to 1426 MW (87%), by 2022. Globally, 16.6% generation of hydroelectricity, 1,079 GW production, in 2015 will be increased to 1,473 GW by 2040 as projected. Although, hydropower is considered as a renewable clean energy, dam closure, influence within the downstream river and connected ecosystems have consequent impacts on hydropower production. Nepal's topography offered more RoR types of hydropower and has more risk of landslide, flooding, GLOFs, LDOFs, and flash floods. Despite, Nepal contributes 0.027% of total global Green House Gas (GHG) emissions; Nepal has focused on renewable energy, hydropower production, targeting 12000 MW by 2030 to fulfill its growing demand of 11,500 MW. Consequent development of clean energy, GHG reduction, single Bhotekoshi hydropower can reduce 160092 tons CO_2/year. The energy-related CO_2 emissions increased 43.2 billion metric tons by 2040 globally, which can be reduced through promotion of clean energy.

Keywords: Water Resources, Hydropower Development, Bhotekoshi Hydropower, Renewable Energy, GHG Reduction, Downstream Impacts

1. Introduction

Nepal has endowed high potential of renewable water resources, possessing about 2.27% of the world's fresh water resources [1]. Most of the rivers flowing from Nepal Himalayas covers

818,500 ha land area equivalent to 5%, out of the total surface area of the country [2]. In total, Nepal possesses 6000 rivers including rivulets and tributaries in totaling of about 45,000 km in length and covering an area of 395,000 ha (48%) [3] and offering dimensional uses including hydropower development [4]. There are 33 rivers having their drainage areas exceeding 1000 km^2 [4] and all the rivers in Nepal comprise the total drainage area of about 194,471 km^2 and the rest in China and India. The annual average discharge of the Nepalese rivers is about 7124 m^3/s including the total basin area and about 5479 m^3/s excluding the area outside of Nepal [5].

Nepal's river has a storage capacity of 202,000 million m^3, which includes about 74% amount from three major rivers, Koshi, Gandaki, and Karnali. Geographically, perennial nature of rivers estimated an annual runoff accounting up to 170 billion m^3 that flows from steep gradient and rugged topography and estimated 45,610 MW, feasible for hydropower generation which is equivalent to 50% of the total theoretical potential of 83, 290 MW [6]. The hydropower system is dominated by run-of-river schemes in Nepal while storage schemes have been benefited to control flood, provide irrigation facility, drinking water supply, navigation, recreation, tourism, aquaculture, and generate revenue [7]. From the beginning, hydropower is used as the natural water-cycle-based renewable energy, which is reliable, most mature, and cost-effective technology of power generation [8]. The discharge and river flow depends on the catchment area, rainfall pattern, and the volume of the water estimates the mechanical energy produced by the falling or flowing water called hydropower.

The source of energy shares from a conventional source in Nepal is 87% as a significant share of electricity and renewable energy with 56.1% households have access of electricity [9]. Hydropower is the main source of energy in Nepal, nearly 90% installed capacity and 90% generation of electricity. The country status report showed that Nepal's energy sources supplied mostly from traditional resources such as firewood (75%), petroleum products (9.24%), animal waste (5.74%), agricultural residue (3.53%), electricity (1.47%), and other renewable resources (0.48%) [10]. Nepal Electricity Authority (NEA) has a total installed capacity of about 746 MW [11] and 26 MW operating from mini and microhydropower plants in the hills and mountains of Nepal [12]. There is a significant energy deficit due to the poor economic and instable government to continue the electricity supply. However, the country has three strategic considerations for exploring large-scale hydropower like storage types of projects, to fulfill country's required demands through installation of medium-sized projects and finally small hydropower projects targeting to fulfill demand of the local communities.

Hydropower is an environment friendly source of energy with no pollution emitting in air or in land, and it also the most efficient method to all. Thus, traditionally hydropower has been considered environment friendly that it represents a clean and renewable energy source.

The multipurpose use of water as fresh water, agricultural, industrial, household, recreational, and environmental and power generation, Nepal's water demand is increasing day by day. Nepal has built several dams for hydroelectricity and irrigation purposes [13]. The human influence with population growth, affluence increase, business activity expansion, and rapid urbanization persuade climate change issue seriously resulting depletion of aquifers and health effects with increasing water pollution. The country's fragile environment consisting rugged topography, monsoon climate, juvenile geology results high rates of runoff, erosion,

http://dx.doi.org/10.5772/66253

landslide, sedimentation, and flooding. The climate change problems and issues are related to the resource impacts concerning the geographical and ecological characteristic of the country. The water resource impact evaluation is challenging toward water availability, quality and stream flow, and sensitive to changes in temperature and precipitation [14]. According to Ref. [15], "inflow of precipitation in the basin depends on the upstream rainfall and snowfall" and affects river regimes in different environmental factors at various timescales like seasonal, monthly, daily, and hourly. Similarly, large-scale hydropower projects may have greater economic and environmental implications. However, the impacts on specific ecosystem of each hydropower depend on the size and flow rate of the river or tributary, climatic, and habitat conditions, type, size, design, and operation of the project, and nature of cumulative impacts that occur upstream or downstream of the river.

2. Material and methods

2.1. Location

Nepal is a land locked and least developed country, occupying 147,181 sq. km total land area, lies between 26°22" and 30°27" N latitudes and 80°04" and 88° 12" E longitudes with north to south of 193 km width and 885 km average stretch from east to west [16]. The country's varying topographic and climatic conditions from south tropical to alpine in north with 80% of the annual rainfall in summer season, distributing both from north-south and east-west directions. Three major river systems originate from across the Himalayan range divided in Nepal from east to west the Koshi, Gandaki, and Karnali River, major tributaries of the Ganges river in northern India.

2.2. Upper Bhote Koshi hydropower: a case study.

The Upper Bhote Koshi Hydroelectric Project (HEP) is a run-of-the-river scheme constructed on the Bhote Koshi river, a tributary of the Sun Koshi river, in Sindhupalchowk district of central Nepal. It is located approximately 110 km northeast of Kathmandu near the Sino-Nepal border; the Project has a head work with a side intake approximately 500 m downstream of the confluence of the Bhote Koshi river and the Jung Khola near Tatopani VDC (**Figure 1**). Water diverted from the side intake is conveyed to a surface powerhouse through a headrace conveyance system comprising a surface desanding basin located adjacent to the weir, a 3.3-km-long headrace tunnel, a restricted-orifice-type surge tank, and a 450-m-long, 2.8-m diameter steel penstock. The Project has an installed capacity of 45 MW, with two turbines/generator units; however, the project is generating 36 MW as per Power Purchase Agreement (PPA) according to the Nepal Electricity Authority (NEA). The Plant was synchronized to the NEA grid on January 3rd, 2001, and commercial operation started on January 24th, 2001. The plants salient generation features are as follows:

- Total generation: 224.970 GWh, 85.2% deemed max.

- 91.5% of estimated average annual generation, plant outages 5.639 GWh (97.7% availability)

- NEA outages 11.046 GWh (95.4% availability)
- Total plant and NEA outages 16.685 GWh (93.1% availability).

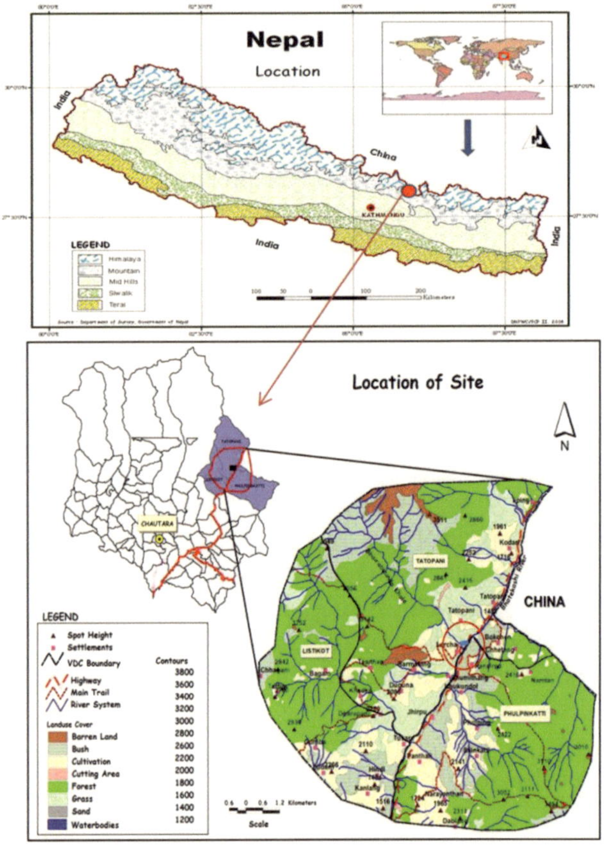

Figure 1. Location of UBHP: a case study area.

2.3. Data

All data used in this chapter were obtained from baseline studies, research, plan and policies, sectoral guidelines and strategies, case studies, and reports. The collected data were compiled, crosschecked, and analyzed from history of hydropower development to current scenario consisting impacts and implications as well as their planning, statistical analysis/downscaling, computational simulations and development of indicators, modeling and projections.

2.3.1. Policies and legal instruments

The policy and legal instruments relevant to hydropower development were reviewed and presented considering the energy issue and analyzed policy level impacts and implications of Nepal's HEPs.

2.3.2. Technical review, Analysis, Projection and Modeling

The relevant data were collected from various sources such as Government of Nepal/ Water Energy Commission Secretariat (GoN/WECS), Department of Electricity Development (DoED), Nepal Electricity Authority (NEA), Department of Hydrology and Meteorology (DoHM), and other hydropower developing agencies. To analyze the baseline scenario of the hydropower project at the national and global level, collected data were computed, projected, and assessed including web-based information. For hydropower generation, analytical methods and procedures engineering toolbox were used to determine and crosschecked the potential capacity of water resources, available power, efficiency, river flow, power generation capacity, hydroelectric power, and hydroelectric energy. The measurement of water discharge has quantifed with comparison of hydrological data and modeling, impacts and implications on hydrology, and flow regime including climate change impact modeling and projections as per the Intergovernmental Panel on Climate Change (IPCC) guidelines. For case comparison, existing overview of the hydropower development, climate change issues, impacts and implications including future scenario with GHG emission and reduction has synthesized from different hydropower's in Nepal, emphasizing details of upper Bhote Koshi hydropower project.

3. Hydropower baseline and electricity generation

The electricity generated by hydropower was 16.6% of the total world's electricity, which is 70% of all renewable electricity [17] and about 3.1% projected to be increased yearly for the next 25 years. The increasing rate of global population by around 1.5 billion projected to be 8.8 billion by 2035 needs 34% increase in energy consumption between 2014 and 2035. The total 549 quadrillion British thermal units (Btu) of world's energy consumption in 2012, would increase up to 815 quadrillion British thermal units (Btu) in 2040 with an increasing ratio of 48% [18]. The average generation capacity of hydro, geothermal, and biomass electricity has fluctuated around 42–43% and the average capacity utilization rate will be 43% for the period of 2016–2040 [19]. The generating capacity of renewable sources of energy like hydropower, geothermal, and biomass 1079, 14, and 52 GW in 2015 will considerably increase globally to 1473, 132, and 275 GW, respectively, as projected for 2040 [20]. Among the renewable source of energy in Asia, hydropower has significant potential with installed capacity of 542 and 2204 GW potential.

The history of hydropower development in Nepal began on May 22, 1911 (9 Jestha 1968 BS) by installing 500 kW electricity at Pharping named as Chandra Jyoti. After 25 years, long duration, Prime Minister Dev Shamsher initiated 640 kW, Sundarijal Hydropower plant with a capacity of 900 kW in 1936. Sundarijal, hydroelectricity development in Nepal was once again stalled for decades. Some years later, Morang Hydropower Company established in 1939 and completed construction of third Letang hydropower plant with an installed capacity of 1800 kW in AD 1943 under public-private partnership. The plant supplied electricity to Biratnagar Jute Mill and later destroyed by landslide [21, 22]. Historically, however, Nepal's first bilateral agreements with India were Koshi and Gandak Projects in 1954 and 1959, respectively, exclusively designed to cater for irrigation and flood control in India with small irrigation and hydropower component for Nepal.

During the late 1960s, a hydropower plant was constructed with foreign assistance such as ex-USSR (Panauti-2.4 MW), India (Trisuli-18 MW, Devighat-14.1 MW, Gandak-15 MW, Surajpura-Kosi-20 MW), and China (Sunkoshi 10 MW, built in 1972 as a gift to Nepal from China). The 92 MW Kulekhani Hydropower Plant (I and II) was commissioned in 1982, which is the only project offering seasonal water storage in Nepal. The 144 MW Kali Gandaki A hydropower project, commissioned in 2003 is the biggest hydropower project in Nepal so far. In 2005, a plan to develop Kaligandaki-Nawalpur diversion (multipurpose) project (with about 20 km tunnel) to generate 22 MW of electricity was formulated but it could not be materialized.

The growing demand of electricity was estimated to be 557 MW in 2005 to 1200 MW in 2013. In 2013, there was 733 MW hydropower out of 782 MW installed capacity including NEA that owned 478 MW capacity of hydropower and 255 MW operated from the private sector. Similarly, electricity demand estimated 105 GWh in 2006/2007 and increasing to 678 GWh in 2009/2010, with the temporary peak in 2008/2009 with 745 GWh. Thus, the problem was mainly due to all hydropower plants which are RoR types except the capacity of 92 MW of storage type [23]. The estimated electricity demand growth rate will be 8.34% increasing from current growth demand 4430 GWh annually to the system peak load of 17,400 GWh with the annual growth projection of 3679 MW by 2027 [24].

Particulars	**Fiscal year**				
	2010/2011	**2011/2012**	**2012/2013**	**2013/2014**	**2014/2015***
Production (MW)	697.85	705.57	746	746	782.45
Transmission line (km)	1917.62	1987.36	1987.36	1987.36	1987.36
Customer number	1,854,275	2,053,259	2,599,152	2,713,804	2,789,678
Distribution line (km)	89,108.86	95,815.98	11,4160.40	116,066.64	116,090.64
Available energy (GWH)	3389.27	3858.37	4260.45	3092.47	3228.9
High demand (MW)	946.1	1026.00	1094	1200.98	1291.1
Demand supply gap (MW)	248	320.43	348	454.98	508.65

Note: *Of the first eight months.
Source: Ministry of Energy, 2016.

Table 1. Electricity demand, consumption, production, and physical structures.

Nepal has an estimated 42,000 MW hydropower potential, 100 MW of microhydropower, 2100 MW of solar power for the grid, and 3000 MW of wind power renewable energy commercially exploitable. Nepal's ninth plan addressed to generate 22,000 MW electricity by 2017 and other studies estimated 10,000 MW within 10 years and 17,000 MW by 2030. The policy of government for small hydropower generation from 1 to25 MW has focused on providing incentives to local institutions/organizations and promoting the development of medium hydropower from >25 to 100 MW for national private sectors and seek support for large hydropower projects. The electricity generation in FY 2013/2014 was 746 MW, which increased by 4.98% (782.45 MW) in 8 months of the year and 1987.36 km transmission line has been extended. The electricity consumers have been increased 2,789,678 in the number and 116,090-km-extended line of electricity distribution. The total demand of the electricity 1291.1 MW has limited to just 782.45 MW by the end of the year (**Table 1**).

The hydropower development started from 1911 generating 1.1 MW electricity, before first five year plan and consequent development in the thirteenth plan which targeted to 668 MW hydropower generation in 2018. The growing energy demand has estimated at 11,500 MW by 2030 for moderate GDP growth (5.6%) and higher demand for GDP growth of >8% . Thus, the existing policy and legal arrangement need to be put in a place, considering the present situation for the sustainable development of hydropower for the overall development of the country. As per the WECS energy strategy 2012, clean energy technology (CET) scenario, in which the fossil fuels should be decreased by 20% by 2020 and 30% by 2030. The produced and targeted electricity generation is shown in **Figure 2**.

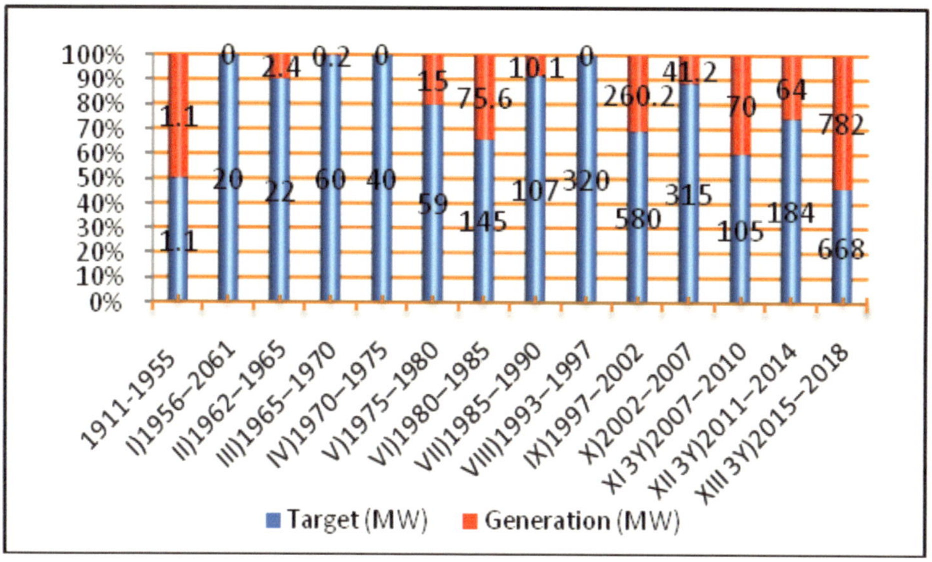

Figure 2. Periodic development of hydropower in Nepal (1911–2016).

3.1. Contribution of hydropower to reduce GHG emissions

The electricity generation and supply become even more important in combination with increasing shares of renewable energies with the low-carbon future envisioned in the Paris agreement. The climate meeting held in Paris in December 2015 (COP21) has superseded global efforts toward embarking upon climate change as the largest source of greenhouse-gas (GHG) emission from energy sector and CC conference (7-18 Nov. 2016, Marrakech, Morocco)

Conference of the Parties (COP 22), agenda for sustainable development goals to reduce greenhouse gas emissions and foster adaptation efforts. The climate leaders of 150 countries representing 90% of global economic activity and 90% of energy-related GHG emissions have submitted a commitment to reduce emissions. The measures submitted by 40% were related to targeting increased renewable energy and one-third submitted suggested improvised energy efficiency [25]. The G7 countries committed in their Declaration on Climate Change to strive "for a transformation of the energy sectors by 2050" and to "accelerate access to renewable energy in Africa and developing countries in other regions[1].

Nepal is a country with least 0.027% of the total global GHG emissions. The total GHG emission recorded in 2008 was 30,011 CO_2-eq Gt, which was found to be increased from 124,541 CO_2-eq Gt GHG emissions in 2000. Considering the country as a party of United Nations Framework Convention on Climate Change (UNFCCC), Nepal limits temperature below 2°C leading to 1.5°C above preindustrial levels. According to the Economic Assessment study of climate change in 2013, the direct cost of current climate variability and extreme events in key sectors agriculture, hydropower, and water-induced disasters has estimated which is equivalent to 1.5–2% of current GDP/year (approximately USD 270–360 million/year in 2013 prices). The model projected for hydropower in lower dry season flows leads lower energy availability, while 2800 MW energy production by 2050 estimated cost will increase by USD 2.6 billion (present value). The economic costs of climate change in hydropower, agriculture, and water-induced disasters could be 2–3% of current GDP/year by midcentury. Thus, the generation of renewable energy like hydropower may reduce the emission trend of the GHG. As suggested in Ref. [25] some areas of Nepal are highly vulnerable in terms of landslide, GLOF and flooding events which can be predict and mitigated through climate resilience strategies and adaptation.

4. Policy and legal instruments in hydropower sector

Nepal's liberal foreign investment policy attracted donor's assistance in hydropower sector since the 1980s. Among the areas of investment in industry manufacturing, services, tourism, construction, agriculture, minerals, and energy, hydropower sector has a widespread investment opportunity due to liberal policy and environmental friendly enacted legal instruments. The government has prioritized the hydropower sector for foreign and domestic investment. As a result, the hydropower sector has received 46% of total Foreign Direct Investment (FDI) in Nepal, investing commitment for Rs 87.56 billion, out of the total FDI commitment of Rs 190 billion received by the end of 9 month in 2016. The Constitution of the Kingdom of Nepal 2015 (2072 BS) has emphasized the development of energy and protection of natural resources. Nepal is governed as per the new Constitution of Nepal, which came into effect on Sept 20, 2015, replacing the Interim Constitution of 2007.

[1] The 21st annual session of the Conference of the Parties (COP) to the UN Framework Convention on Climate Change (UNFCCC). International energy efficiency initiatives were undertaken during 2015 by several United Nations (UN) Regional Commissions, the UN Development Programme, the Global Environment Facility (GEF), the World Bank, the IEA, the EU-GCC Clean Energy Network and the Green Climate Fund of the UN Framework Convention on Climate Change (UNFCCC).

http://dx.doi.org/10.5772/66253

The prevailing policies, acts, and regulations are: Hydropower Development Policy 1992, Water Resources Act 1992, and Electricity Act 1992. For management and development of electricity services, Electricity Act, 1992, Electricity Rules, 1993, and Electricity Tariff Fixation Rules, 1993 are the major legal regimes. The major objectives of the Hydropower Development Policy 1992 were to involve private investment in hydropower generation: in order to fulfill these objectives, concept of BOOT (Build, Operate, Own, and Transfer) in developing hydro projects were introduced. The Hydropower Development Policy 1992, supported by the Electricity Act, 1992, provides incentives to develop hydropower in Nepal.

Similarly, Foreign Investment and One Industrial Policy, 1992, Industrial Enterprises Act 1992, Foreign Investment and Technology Transfer Act, 1992, Environment Protection Act and Environment Protection Regulation, 1997 were enacted to strengthen the sector together with protecting the natural environment. The promulgated Ninth Five-Year Plan (1997–2002) of the government had included institutional reforms to attract private sector in power generation and distribution and various programs such as generation and supply of electricity, power transmission, system strengthening, feasibility study, and design for rural electrification. Likewise, Tenth Five-Year Plan Period (2002–2007) emphasized small, medium, large, and reservoir types of hydropower construction and formulated Water Resources Strategy, 2002 and National Water Plan, 2005. The plan intended to promote integrated development of water resources involving private and public sectors with emphasis on rural electrification and control of unauthorized leakage of electricity. The second Three-Year Plan (2010/2011–2012/2013 (2067/2068–2069/2070) has given special encouragement to not only public sectors but also for private sectors with public-private partnership aspects. Considering the existing legal regime, the total 733.557 MW hydropower has produced up to the year 2014/2015 including 255.647 MW from the private sector. There are 83 projects equivalent to 1521.28 MW installed capacity that are in the construction phase and 33 HEP with 532.542 MW installed capacity that are in different development phases.

Nepal's concern toward conservation and sustainable economic development has signatory of more than 20 international treaty and conventions such as Biological Diversity (CBD) and UN Framework Convention on Climate Change (UNFCCC) in 1992, UN Convention on Combating Desertification (UNCCD), 1994 and Kyoto Protocol in 1997. The climate change issues raised in Nepal formed Climate Change Council in 1999 and adopted Climate Change Policy in 2011 with its main aim to improve socioeconomic development with improvement of livelihoods by mitigating and adapting adverse impacts and adoption of low-carbon emission-based development.

Nepal's Climate Change Policy (2011) has envisaged protection of environment and sustainable human development by promoting the use of clean energy, reducing GHG emissions, enhancing the climate adaptation, and resilience capacity of local communities. The emissions of carbon threaten basic elements of society like water, food production, health, and the environment imposing a huge social cost as anywhere from $8 per ton to as high as $100 per ton of CO_2. Fossil fuel combustion for transportation and electricity generation are the main source of CO_2 generation, contributing more than 50% of the emissions, and generation of electricity with thermal power plants contributes 66% the world's electric generation capacity.

However, the hydropower represents only 20% of the electricity generation capacity of the world, which emits 35–70 times less GHGs per TWh than thermal power plants [26]. To achieve energy efficiency and energy security, the hydropower projects aim to access affordable and reliable energy service and reduce high carbon emission replacing petroleum products to the extent possible in transport, industry, and household sectors.

The main objective of Thirteenth Three-Year Plan (2013–2016) on Water Sector [12] is to upgrade living standards of the people with their socioeconomic development and reduction of poverty line as below 18% [12]. The plan emphasized to upgrade Nepal accessing people toward drinking water from 85 to 96.25%; sanitation from 62 to 90.5%, and grid-connected electricity generation from 758 to 1426 MW accessing electricity from 67.3 to 87% and ranked Nepal as a developed country by 2022.

5. Impacts and implications

5.1. Adverse impacts of hydropower development

Although hydropower is known as clean energy, it has many associated environmental impacts and implications, like first-order abiotic and biotic impact occurs simultaneously with dam closure and influence within the downstream river and connected ecosystems (changes in flow, water quality, and sediment load). The second-order modified the first-order impacts changing channel with downstream ecosystem and primary productivity in long terms. The third-order impacts are biotic, changes resulting from the integrated effect of all the first- and second-order changes, including the impact on species close to the top of the food chain changes in invertebrate communities and fish, birds, and mammals [27]. Similarly, impacts of hydropower development defined as serious and significantly leading to increasing poverty, social dislocation, and loss of terrestrial and aquatic biodiversity especially dealing to fish resources [27–31].

The river basins in Nepal have extensive pressure due to developmental and demographic growth creating consequent adverse effects on aquatic biodiversity especially the native fish fauna. The greatest threat occurs to the migratory fishes, as their migratory path is obstructed by hydropower structures and fish species are depleting [32–35]. The hydropower generation changes the river system by obstructing habitat ground of the fish species. Fish migration from the downstream barrier to the upstream barrier created dam and passing through turbines, spillways, or in the diversion subjected to injury by physical contact, pressure change, shears force, or eddies. The nature of the river bottom will change, the water quality may also change [36, 37]. A study on the impact of dams on different rivers in Nepal indicated its potential to alter the health and integrity of the rivers with effects being more serious downstream [38].

Adverse impacts of hydropower projects poses project-specific impacts related to project design, size, types, location including physical, biological socioeconomic, and cultural environmental characteristics [15]. There may have fewer adverse impacts on RoR types of projects with high head and small reservoirs with much smaller foot print than the large reservoirs. However, additional benefits found some large reservoirs like flood control, irrigation, aquiculture, and recreation/tourism opportunities [39–44]. Similarly, summarizing the previous studies and research findings [10] has observed and analyzed key environmental challenges and impacts of the hydropower sector in Nepal such as landslide, sediments load, loss of biodiversity, declining fish species, alternation of hydrological pattern, change in river morphology, water pollution, and other associate social impacts.

The detailed study of Bhote Koshi HEP revels that the project impact on forests of project and the influenced area cannot recover the natural condition of the local environment but the effective implementation of EIA mitigation measures could reduce the impacts [45]. The identified monitoring indicators and parameters during construction, operation, and implementation phases of upper Bhote Koshi regarding: (i) physical-air quality, noise level, vibration, water quality, erosion and stability of slope, (ii) biological-forest and vegetation, wildlife, and biodiversity and fisheries, and (iii) socioeconomic environment, all these parameters are limited on natural disasters like, flooding, landslide, erosion, and possible risk of GLOF only.

5.2. Climate change and vulnerability

Nepal is one of the most vulnerable countries to climate change, water-induced disasters and hydrometeorological extreme events such as droughts, storms, floods, inundation, landslides, debris flow, soil erosion, and avalanches [46]. According to previous report, 22 districts are highly vulnerable in terms of landslide-prone areas, 12 districts to GLOF, and 9 for flooding. National communication report 2015 described an increasing trend of climate change from the energy sector. In the case of hydropower, the model projected lower dry season flows and thus lower energy availability while 2800 MW energy production by 2050 with an increased cost of USD 2.6 billion (present value) was estimated. The economic costs of climate change in hydropower, agriculture, and water-induced disasters could be 2–3% of current GDP/year by midcentury.

Internationally determined contributions [47], acknowledge that the high vulnerability of Nepal's energy sector to climate change, and identified this as a crucial dimension of the country's overall vulnerability to climate change, and as a critical threat to the economic well-being, livelihoods, and energy security of the Nepalese population. Nepal's hydropower plants and indeed its entire energy system are already vulnerable to extreme weather events, as made clear by the NEA/WECES Energy Status Reports. However, for implementation of INDC, Nepal will contribute to global efforts to reduce GHG emissions for life-support systems adapting and build climate resilience development to mitigate climate change impacts. The renewable energy demand needs to be integrated in other sectoral policies [48].

Nepal's mountainous and challenging topography and socioeconomic conditions (ranks 145 on the Human Development Index, nearly one-fourth of its population live below poverty line) make it a highly vulnerable country to climate change. Nepal has experienced changes in temperature and mean precipitation. The country, with the exception of some isolated pockets, has become warmer. Data on trends from 1975 to 2005 showed 0.06°C increase in temperature annually whereas mean rainfall has significantly decreased to an average of 3.7 mm (−3.2%) per month per decade. Under various climate change scenarios, mean annual temperatures are projected to increase between 1.3 and 3.8°C by the 2060s and 1.8 and 5.8°C by the 2090s. Annual precipitation is projected to reduce in a range of 10–20% across the country.

These vulnerabilities are being exacerbated by climate change. As detailed in Nepal's Second National Communication [49], Nepal's hydropower plants are highly vulnerable to the projected impacts of climate change as they depend upon river basins fed by glacial meltwater and snowmelt. The challenging environment in Nepal, Himalayan glaciers are receding faster than the other glaciers in the world [50]. The total 3252 glaciers in Nepal covering a total area of 5312 km^2 studied in 2001, whereas 3808 glaciers covering an area of 4212 km^2 were reported in the study of 2010 [51]. Most climate models predict significant changes in the dynamics of mountain glaciers, snowmelt, and precipitation as the climate warms. The International Commission on Large Dams [52] has already emphasized the urgent need to adapt older dams to cope with the impacts of climate change. At the same time, Nepal's poverty reduction strategy emphasizes the importance of increasing availability of affordable energy and using Nepal's abundant hydropower resources to promote economic growth and development.

5.3. Natural disaster

Nepal (relief) Natural Calamities Act, 1992, concerning most devastating disasters, earthquakes since 1934, 1980, 1988–2015 and summer flooding events of 1993, 2008, 2013, and 2014. The another great event of 7.8 magnitude earthquake struck Nepal on 25 April 2015 (11:56 am local time) epicenter in Barpak Village of Gorkha district, 81 km northwest from the capital city of Kathmandu [53]. The earthquake directly affected 14 districts with human and livestock casualties, infrastructure, and physical property. The Gorkha earthquake 2015, damaged buildings and infrastructures like road, hydropower projects, residential, and school buildings. Burial of the entire Langtang village, blocking of Kaligandaki river, damages to hydropower projects and damages to Araniko and Rasuwagadhi highways connecting to China were the major noticeable events due to the earthquake-induced landslide.

Figure 3. Flooding washed way Bhotekoshi Hydropower Project, July 6, 2016 4:11.

In addition, numerous stretches of the feeder roads and trails were also damaged [54]. Bhote Koshi hydropower project was affected by earthquake-induced landslides damaging penstock

http://dx.doi.org/10.5772/66253

stock and powerhouse sites destruction. Similarly, landslide damaged penstock alignment of Sanima, powerhouse site of Aankhu Khola, Chaku Khola and middle Bhote Koshi hydropower projects. A year later of earthquakes struck, landslide dam burst in Tibet has caused heavy destruction at downstream sections of Bhotekoshi river in July 2016 (**Figure 3**).

Similarly, the Jure landslide occurred in August 2014 brought down a whole mountainside and blocked the river. Landslides that hit the entire Jure bazaar in Mankha of Sindhupalchok, killing an unknown number of people, went on to block the course of Sunkoshi river. The river blockage, on the other hand, has imposed a threat of outburst floods, threatening thousands of people living downstream, and inundated at upstream. Nearly 10% of the nation's hydropower capacity, some 67 MW, was severed by the landslide, submerging a 5 MW power plant and disconnection of the power supply with 45 MW Bhotekosi hydropower and 10 MW Sunkosi hydropower and washed out over 400 houses, killing over 200 people.

5.3.1. Glacial lake outburst floods (GLOF)

In region of the great Himalaya, 17% of the area covered by glacier and ice cover and nearly 113,000 km^2, the total area covered by glaciers and permafrost outside the polar region. The Himalayan region has 35,000 km^2 of glaciers, and a total of 3700 km^3 ice reserve. The Himalayan region is the source of nine largest rivers in Asia where over 1.3 billion people lived. The warming in Nepal and Tibet increased progressively within a range of 0.2–0.6°C per decade between 1951 and 2001, particularly during autumn and winter. Over the past 30 years, the length of the growing season has increased by almost 15 days. The Himalayas glacier melt has increased the frequency and intensity of risks like floods, avalanches, failure of moraine-dammed lakes, and water regime affects.

Figure 4. Impacts of GLOF of hydropower UBHP (45 MW), July 2016.

Thus, the breaks of dams devastating debris flows occur, internationally called GLOFs. As several downstream hydropower are in risk in Nepal, recently July 16, 2016 UBHP (36 MW) was destroyed by GLOF generated from Nyalm Tibet (**Figure 4**). Previous recorded events of NEA, Namche HEP swept away in 1985 through Dig Tsho GLOF, and huge flood struck on 27 August 2008 in 4 MW Khudi HEP.

5.4. Policy level implications in hydropower sector

The concerned policies like Water Resources Act, 2049, Electricity Act, 2049, Water Resource Regulation, 2050, Electricity Regulation, 2050 in the hydropower sector have not described threshold criteria in terms of environmental sensitivity. Previously, 5-MW threshold criteria of EIA guideline 1993 repeatedly addressed in the Environment Protection Act and Rules, 1997 for the environmental assessment process. This criterion had raised issues in a licensing process and well as environmental assessment process of hydropower generation. The Water Resources Development Policy, 2001, has a provision to promote an integration of environmental aspects during the development of water resource sector [55]. The policy urges to ensure minimum release of 10% discharge or more as recommended by the EIA study during the construction and operation of hydropower projects, and encourages the private developers to acquire necessary land for the project by themselves. These criteria are not applicable for every project considering the size, location, and sensitivity. Owing to the policy, electricity supply is limited to 43.6% for urban population (2009) and only 8% access in rural areas. As a result, Nepal has limited access and low-level electricity consumption among other developing countries [23]. The policy level document of the sector has not defined high dam impacts and mitigation measures. All policy is conventional and needs to be reformulated in the present context.

5.4.1. Policy level implications in EIA process

In hydropower projects, Ministry of Water Resource (MOWR) is responsible for compliance monitoring and role of environmental assessment facilitator for sectoral agencies. The Ministry has a lead role with project proponent and reviewer's role for monitoring of compliance and monitoring from the Ministry of Environment as an EIA approval agency. The Ministry of Environment (MoE), a leading agency of all EIA approval process, monitoring is inactive due to lack of trained human resources as well as guided policy. The major policy level implications regarding protected species of government, CITES Appendices and IUCN red list category, maintaining 40% forest area, downstream release of 10% water in HEPs, assessed impacts were found significant even if a single tree to be felled during project implementation. Consequently, downstream release of 10% water in hydropower projects and significant impact of tree felled has observed policy level constraints. Similarly, it has assessed that the provisions related to prediction, monitoring, and evaluation of impacts has not fully considered in EIA studies and its implementation as per EPR 1997.

EIA reports generally recommended mitigation measures for adverse impacts of the commenced power projects but trend remains limited to implementation of major issues and impacts. Not enough considerations have been made in the issue of ambient air pollution,

http://dx.doi.org/10.5772/66253

noise, and vibration control specifications as well as protected flora and fauna Contractor's responsibility for arranging to provide awareness for environmental protection is not addressed in the tender document. Some of the mitigation measures recommended in EIA report were seen irrelevant during operation as serious shortcoming in EIA qualitative data (e.g., species lists, distribution, and habitats) identified previously in some studies [56–59]. To produce cheaper hydro energy, national capabilities need to be strengthened, although there are some legal issues for the development of hydropower such as nonspecificity of water rights and ownership; lack of subordinate enabling legislation; lack of harmony among related legislation; and lack of adequate legal provisions to encourage private sector participation in multipurpose projects [60].

5.5. Challenges and opportunities

The major thrust for hydropower development in Nepal is the financial investment and risk of natural catastrophe. On the other hand, the environmental and social impacts are linked to the development of hydropower projects which obey misrepresented and this delays the process, or in some cases stops the development of hydropower project. The World Bank had withdrawn the Arun III Project in 1995 in Nepal [31], concerning environmental and social impacts of this project and the government instability might be the reason behind it. In 2007, NEA sold 40.7% of its electricity to domestic users, 38% to industries, 6.6% to commercial and rest to noncommercial including the agricultural sector at 79 kWh [61]. Nepal's per capita consumption of electricity is one of the lowest in the world; however, some countries like Singapore has the per capita 6500 kWh consumption. Nepal's dependence on technical for planning, designing, and constructing water development projects and financial support for large-scale projects slow down the hydropower development process. The Power Investment Summit organized by the Energy Development Council (EDC) 2016 concluded that Nepal requires $20 billion to develop 10,000 MW on grid hydropower projects in the next 10 years. Investment of $5 billion needed for high-voltage transmission line projects to be completed within 2035. The investment opportunities on Budhigandaki 1200 MW, Nalsingad 410 MW, Tamor 762 MW, Andhikhola 180 MW, Tamakoshi V 87 MW, Upper Tamor 415 MW, Tamakoshi III 650 MW, and Thuli Bheri 530 MW projects have been identified issuing a statement by the EDC.

Landslide dam outburst floods (LDOFs) is one of the major challenges for hydropower development in Nepal due to its rugged topography, susceptible to landslides, very high relief, and intense precipitation during the monsoon period, for example, UBHEP event of July 2016 learnt more about it. As the glacier melt, water volume initially increases with an increase of temperature and hydropower potential generally increases accordingly, as it largely depends on the lean season flows. Nepal's hydropower generation generally follows the pattern of dry season flows [62]. Over 90% of Nepal's existing hydropower plants are the runoff river type, which are generally designed based on the dry season flows. These power plants have been already facing the problem of water shortages during dry seasons and generating only about 30% of the total installed capacity in dry months. The problem will be further exacerbated during dry season by the reduced snow and glacier-melt contribution in the future [63].

However, there is a new opportunity for hydropower from climate change is to increase-installed capacity because increasing precipitation and storage projects help to adapt climate change for local people [64].

The flooding events like GLOFs, LDOFs, and flash floods increase sediment or debris flow on the reservoir that may loss production capacity. In July 1993, Kulekhani project had turn out such event causing high sediment and debris inflow to the reservoir because of flood. Similarly, in 2014, landslide occurred during monsoon which badly hit United Modi Project 10 MW, and damaged the concrete cover slab of hundreds of meters of the headrace tunnel and canal covered by debris. Out of the total available energy in 2012, only less than 1% was from the thermal sources and about 80% was from domestic hydropower plants [65]. Any decrease in the river flows during dry season due to the decreased glacier ice reserve and decreased flows of the rain-fed rivers as projected will further deteriorate the electricity generation.

Nepal's energy market is dependent on hydropower generation. The power demand for Nepal was 470 MW in 2002–2003 with available 2261 GWh, out of which 2107 GWh energy system within NEA was met by hydropower generation (93%). Despite having a potential to generate 43,000 MW of electricity, Nepal's installed hydropower capacity is just 787 MW, which is less than half of the demand. Even at a high growth scenario of about 12% per annum, peak power demand will reach only 3400 MW and energy requirement 16,000 GWh in 2020. Thus, even in the foreseeable future Nepal's electricity needs will still be a small percentage of its realizable hydropower potential. Thus, in order for Nepal to exploit its hydropower potential in a substantive way, it has to look for an export market where there is a demand for such power (Nepal's Electricity Act, 1992, provides for export of electricity generated by a developer to foreign country by entering into an agreement with the government. The developer will have to pay export duty as determined in such agreement).

6. Mitigation and adaptation

Nepal has planned to produce 12,000 MW clean energy by 2030 including 4000 MW hydro-electricity by 2020, 2100 MW solar energy, 220 MW bioenergy by 2030, and 50 MW of electricity from small and micro-hydropower plants. Nepal government has a strategy to maintain 40% of the total forest area of the country promoting conservation of biodiversity, resilience infrastructure, a forestation in public, and private lands. The ecosystem of ecoregions of the country supports to be managed endorsing sustainable management of forests, enhance capacity of local communities' adaptation and resilience, widen carbon storage through sustainable forest management, and reduce carbon emissions.

The conventional or ongoing approach to determine the dam and reservoirs capacity during planning and design phases on the basis of limited data and estimation of sediment discharge could be improved by considering seismic design through ICOLD Bulletin 148 (2010), guidelines for the study of hydropower projects [66]. Likewise, GLOF risks are difficult to predict but proper assessment of glacial hazards in upstream catchments, reviewed, and monitored regularly, preferably every 5 years may help for preparedness and appropriate strategies for mitigation of the probable risk [67]. For such events, Nepal has adopted national policies, strategies, and program such as Sustainable Agenda for Nepal (2003), Water Resources Strategy (2002), National Water Plan (2005), and Disaster Risk Reduction Strategy (2009) which can be the best mitigation measures for HEP development.

To evaluate possible risks or failure and downstream hazards all actions such as early warning system, emergency action planning, dam-break hazard analyses, inundation mapping, dam-break flood analyses, and dam-break mechanism practices for emergency action planning are necessary to be taken or embarked. The key contribution of afforestation activities in tropical zones avoiding deforestation, or REDD (Reducing Emissions from Deforestation and Forest Degradation), Kyoto Protocol, Clean Development Mechanism (CDM) negotiations are underway for the future international climate agreement. Similarly, internally emerged effective and environmentally, economically and socially adaptive ecosystem-based approach can be resilient the hydropower infrastructure and reservoirs to cope with the impact of climate change. The storage hydropower brings climate adaptation services like flood protection, source of water supply for agricultural, industrial, urban, and/or environmental purposes. Nepal's sustainable goal 2016–2030 emphasized major objectives to access quantity and quality-based electricity supply with efficiency, safety and convenience of household energy, and promotion of hydropower potential as clean energy in the South Asian region [68]. Thus, there is a great need to protect the ecosystem together with the hydropower development by improving resilient hydropower infrastructure through good planning, design and sitting; construction; operation and maintenance, contingency planning, and restoring ecofriendly environment.

The benefit of hydropower can be analyzed, the example of a case study that the Upper Bhote Koshi plant (45 MW) annual electricity output with 70% Plant Capacity Factor would be 275,940 MWh/year. To determine expected project emissions of hydropower annual CO_2 emissions (tons CO_2) = zero project emissions. The annual baseline emissions reduction for UBHEP only will be 160,092 tons CO_2/year. Since 2001–2016, Bhote Koshi HEP alone has reduced approx. 2,49,1380 tonsCO_2.

7. Conclusion

The global energy agenda for the seventh successive year in 2016 renewable energies is the high impact issue for the feature, with a strong perception to reduced uncertainty, as a top action for priority agenda in the energy sector, to build on the importance of developments around both climatic framework and innovation. The countries determinant commitment in COP21, toward an increase in the anticipation for further scaling up of renewable energies, reinforcing the reduced level of uncertainty as large hydro is understandably an issue closely aligned to localized resource availability and sustainable development 2030s goals for benefiting and adapting climate change of COP 22s efforts. Hydropower development is the top priority action of global energy leaders such as Latin America, followed in priority for those in Asia and Africa who perceive the issue with similar low uncertainty but a relatively low degree of impact. All countries have energy efficiency priority in policy objectives bearing in mind its multiples benefits including prominent role on climate change issues/GHG reductions. In this context, Asia has estimated the largest remaining unutilized potential of hydropower at 7195 TWh/year, making it the likely leading market for future development [69] and support to reduce global GHG emission in the near future.

The hydropower potential in Nepal depends on the 6000 or more rivers flowing from mountains to hills and plains, although Nepal is divided into five geographic regions namely Terai plain, Siwalik Hills, Middle Mountains, High Mountains, and the High Himalayas. The world's water coverage of 97% seawater and 3% fresh water, Nepal constitutes nearly 5000 lakes; 1380 reservoirs; and 5183 village ponds including 3808 glaciers with a total area of 4212 km^2, and 1466 glacial lakes with an area of 64.75 km^2. Among these glacial lakes, 20 lakes were identified as a potential risk of glacial lake outburst floods [70]. Despite the natural beauty of Nepal with its unique topography, ecological regions, geography, and enormous source of water potential, Nepal is the poorest countries in the world wherever electricity infrastructure is heavily reliant for hydroelectricity power generation.

Despite the high potential of hydropower potential, Nepal's low economy and slow GDP growth rate in combination with environmental and socioeconomic constraints, effective implementation of existing policy and political stability may support to reach the sustainable development goals of the county. Nepal however ratified the UNFCCC, the Convention to Combat Desertification (UNCCD), and the Convention on Biodiversity (UNCBD), the most critical impacts of climate change consisting water resources and hydropower generation, stemming from glacier retreat, expansion of glacial lakes, and changes in seasonality and intensity of precipitation through the grid and off grid system. The projected climate change scenarios for Nepal average mean temperature will increase by 1.2 and 3°C as projected by 2050 and 2100. This trend is significantly dangerous for glacier retreat and glacial lakes expansion, making them more prone to GLOF. The GLOFs significant impacts on hydropower, rural livelihoods, and agriculture and storages dams anticipated a climate change impacts which poses environmental conflicts for sustainable hydropower development.

The national and international norms, policies, laws, and regulations for maintenance of ecological integrity are generally not fulfilled as per the case study of UBHP. Maintenance of 40% forest area, downstream release of 10% water, and compensation even for singletree are the significant constraints short out in implementing-related policies. The benefits of hydropower development should be considered with analysis of cost benefit, technical accuracy, and scientific structural design, implementation of relevant policy and legal instruments, strong implementation of EIA mitigation measures might make hydropower sustainable clean energy reducing global warming, and dependency on fossil fuel. The impacts on physical receptors especially downstream flow and change in river morphology have generated major concern as they remain as cumulative impacts. The fluctuation of minimum downstream release was found as the major cause for declining fish diversity. More anthropogenic disturbances and project structures were found creating impact on extinction of wildlife and protected species like birds and mammals. Furthermore, hydropower development has negative environmental and social consequences, although it does not emit air pollution or greenhouse gas emissions. Since 1970, global hydroelectric capacity doubled and dam blocking of rivers degrading water quality, disturbance of aquatic and riparian habitat, migratory fish route blockage, economic impacts on fishery, and displacement of local communities.

Author details

Ramesh Prasad Bhatt

Address all correspondence to: drrameshbhatta@gmail.com

Institute of Ecology and Environment, Budhanilkhantha, Kathmandu, Nepal

References

[1] CBS. Statistical Year Book of Nepal. s.l.: Central Bureau of Statistics, Kathmandu, Nepal, 2005.

[2] Bhandari, B. The Current Status of Wetlands in Nepal, Country Report Presented at the Asian Wetland Symposium . Otsu/Kushiro, Japan: Organized by Ramsar Center, Japan, 14–20 October 1990, 1992.

[3] DOFD. Annual Progress Report. Fisheries Development Directorate, Government of Nepal, Kathmandu, 2003/2004.

[4] WECS. Water Resources of Nepal in the Context of Climate Change. s.l.: Government of Nepal, Water and Energy Commission Secretariat (WECS), Singha Durbar, Kathmandu, Nepal, 2011.

[5] CBS. Statistical Year Book of Nepal 2013. s.l.: Government of Nepal, National Planning Commission Secretariat, Central Bureau of Statistics (CBS), Kathmandu, Nepal, 2013.

[6] Amatya. Updates on Energy Scenario and Energy Balance, s.l.: Bulletin, 1997, 8(1), 1997.

[7] Bhatt, R.P. and S.N. Khanal. Impoundment After Damming the Rivers Change in Flow Regime and Effect in Water Quality of Chilime Hydropower Project in Rasuwa District Nepal. s.l.: Advances in Bioresearch, 2011, 2, 33–39.

[8] Brown, A., S. Müller and Z. Dobrotková. Renewable Energy Markets and Prospects by Technology. International Energy Agency (IEA)/OECD, Paris, 2011.

[9] MOE. A Report on 20 Years Hydropower Development Programme. s.l.: Ministry of Energy (MOE), Kathmandu, Nepal, 2010.

[10] Bhatt R.P. Ecological Impacts and Implications of EIA Studies in Nepal. PhD Thesis, Kathmandu University, Kathmandu, Nepal, 2012.

[11] MoF. Economic Survey. s.l.: Ministry of Finance, Government of Nepal, Kathmandu, Nepal, 2014.

[12] NPC. Approach Paper to Thirteenth Plan (FY 2070/71). National Planning Commission, Kathmandu, 2013.

[13] Jha, J. and R. Chaudhary. Hydropower Development and Fish Diversity: A Paper Presented at the Workshop on Hydropower Dams: Impact on Fish Biodiversity and Mitigation Approached. Kathmandu, s.n., 2003.

[14] WECS. Water Resources of Nepal in the Context of Climate Change. s.l.: Water and Energy Commission Secretariat (WECS), Kathmandu, Nepal, 2011.

[15] Bhatt RP. 2016. Climate Change Impacts and Flow Regime Alternation in Indrawati River Affecting the Fish Diversity, Journal of Environmental Science, Computer Science and Engineering & Technology, JECET; 2016; Sec. A; Vol.5. No.3, 612–639.

[16] NPC. Statistical Pocket Book Nepal 2010. Government of Nepal, National Planning Commission, Central Bureau of Statistics, Kathmandu, Nepal, 2010.

[17] REN, 21. Renewables 2016. Global Status Report. Renewable Energy Policy Network for 21th Century. Renewable Energy Policy Network, REN21 Secretariat, Rue Miollis, Paris France, 2016.

[18] IEO, 2016. International Energy Outlook 2016 (IEO2016). U.S. Energy Information Administration (EIA), Washington, DC, 2016.

[19] BP. Statistical Review of World Energy June 2016. BP p.l.c, St James's Square London, SW1Y 4PD, UK, 2016.

[20] EIA. Annual Energy Outlook 2016. US Department of Energy, US Energy Information Administration, Washington DC, 2016.

[21] Pradhan, P.M.S. Hydropower Development. NTNU, Hydro Lab Pvt. Ltd. Kathmandu, Nepal, 2006.

[22] IPPAN. Hydropower related Policies and Legal provisions. Consensus and Capacity building for Hydropower Projects, IPPAN, Nepal Hydropower Association, Pokhara, 12 April, 2015.

[23] NEA. Annual Report, Fiscal Year (2009/2010). Nepal Electricity Authority (NEA), Kathmandu, Nepal, 2011.

[24] WECS. Energy Synopsis Report. s.l.: Water and Energy Commission Secretariat (WECS), Kathmandu, Nepal, 2010.

[25] MoSTE. National Adaptation Programs of Action (NAPA). Ministry of Science Technology and Environment, Kathmandu, Nepal, 2010.

[26] IAEA. Hydro-Québec (1996). Assessment of Greenhouse Gas Emissions from the Full Energy Chain for Hydropower, Nuclear Power and Other Energy Sources. Working Material: Papers Presented at IAEA Advisory Group Meeting Jointly Organized by Hydro-Québec and IAEA. s.l.: (Montréal, 12–14, March, 1996), Montréal, International Atomic Energy Agency et Hydro Québec, 1996.

[27] McCartney, M.P., Sullivan, C. and Acreman, M.C. Ecosystem impacts of large dams. IUCN, Centre for Ecology and Hydrology, Wallingford, UK, 1999, 75pp.

[28] WB. The World Bank's Experience with Large Dams: A Preliminary Review of Impacts. Profiles of Large Dams (background document). Operations Evaluation Office, The World Bank, Washington, DC, 1996.

[29] WWF. A Place for Dams in the 21st Century? World Wildlife Fund, Gland, Switzerland, 1999, 40 pp.

[30] Scudeller, V.V., F.R. Martins, and G.J. Shepherd. Distribution and Abundance of Arboreal Species in the Atlantic Ombrophilous Dense Forest in Southeastern Brazil. s.l.: Plant Ecology, 2001, 152(2), 185–199.

[31] WCD. Dams and Development: A New Framework for Decision-Making. The Report of the World Commission on Dams.: Earthscan Publications, London, UK and Sterling, VA, 2000. URL: www.dams.org.

[32] Naidu, B.S.K. Environmental Aspects of Chamera Hydroelectric Project (540 MW) in Himachal Pradesh. In: Environmental Impacts of Water Resources Development, National Round Table Discussion. Goel, R.S. (eds.) s.l.: Tata McGraw–Hill Publishing Company Limited, New Delhi, India, 1993, June 4–5, 1993.

[33] Shrestha, K.K. and P.K. Jha. Plant Diversity and Evaluation of Conservation Measures in the Bardia National Park (BNP). A report submitted to World Wildlife Fund Nepal Program, Kathmandu, Nepal, 1997.

[34] Shrestha, J. Taxonomic Revision of Fishes of Nepal. In: Environment and Agriculture: Biodiversity, Agriculture and Pollution in South Asia. Jha, P.K., Baral, S.R., Karmacharya, S.B., Lekhak, H.D., Lacoul, P. and Baniya, C.B. (eds.). Ecological Society of Nepal, Kathmandu, Nepal, 2001, pp. 171–180.

[35] Arya, S.C., K.S. Rao, and S. Shrivastava. Biodiversity and Fishery Potential of Narmada Basin Western Zone (M.P., India) with Special Reference to Fish Conservation. Environment and Agriculture. s.l.: Agriculture and Pollution in South Asia, 2001, pp. 108–112.

[36] Moss, B.l. Ecology of Fresh Waters Man and Medium, Past to Future. s.l.: 3rd edition. Blackwell Publishing, University of Liverpool, UK, 1998, pp. 118–122.

[37] Gutzmer, M.P., J.W. King, D.P. Overhue and Y. Chrisp, (eds.). Fish species – Richness Trends in the Niobrara River, Nebraska, below the Spencer Dam. Transactions of the Nebraska Academy of Sciences, USA, 2002, Vol. 28, pp. 57–62.

[38] Jha, B.R., H. Waidbacher, S. Sharma and M. Straif. Fish Base Study of the Impacts of Dams in Different Rivers of Nepal and its Seasonal Variations. s.l.: Ultra Science, 2007, 19(1), 27–44.

[39] Onta, I.R. Large Dams and Alternatives in Nepal: Experience and Lessons Learnt: A Paper Presented at the World Commission on Dams Regional Consultation. East Consult, Kathmandu, 1998. URL: www.dams.org/submissions/sub_n2.htm.

[40] Garcia, J.C. A Regional Review of the Environmental and Social Impacts of Hydroelectric Projects: Environmental Component. Garcia and Associates, San Anselmo, CA, 1999. (Prepared for the Operations Evaluation Office, Asian Development Bank).

[41] Garcia, J.C. and C.T. Garcia. A Review of Environmental Impacts of Hydroelectric Projects in Asia. Prepared for the HydroVision Conference, Raleigh, NC, 2000.

[42] Pandey, B. Environmental Impacts of Kali Gandaki 'A' Hydroelectric Project on Vegetation Resources in the Dam and Reservoir Area. PhD dissertation. Central Department of Botany, Tribhuvan University, Kathmandu, 2001.

[43] Upadhaya, K.K. and B.C. Shrestha. Project Induced Impacts on Fisheries Resource and their Mitigation Approach in the Kali Gandaki 'A' Hydroelectric Project, Nepal, In Cold Water Fisheries in the Trans-Himalayan Countries, Petr, T., Swar, D.B.T. and Swar, D.B. (eds.). FAO Fisheries Technical Paper, Rome, 2002, 431 pp.

[44] Garcia, J.C., C.T. Garcia, S. Devkota and R.P. Thanju. Resettlement: Lessons Learned at Kali Gandaki A in Nepal. s.l.: Hydro Review Worldwide (HRW) Magazine, 2005, 13(1), 32–37.

[45] Bhatt R.P., Khanal S.N. Vegetation analysis and differences in local environment variables in indrawati hydropower project areas in Nepal. International Research Journal of Plant Science (ISSN: 2141–5447) Vol. 1(4) pp. 083–094, 2010.

[46] NAPA. Nepal National Adaptation Programme of Action (NAPA) Official Document – September 2010. National Planning Commission, Ministry of Environment, Kathmandu, Nepal, 2010.

[47] INDC. UNFCCC Synthesis Report, Intended Nationally Determined Contribution. International Center For Climate Governance, Island of San Giorgio Maggiore, Venice, Italy, 2016.

[48] MoSTE. Climate Change Policy, 2011. Ministry of Science, Technology and Environment, Government of Nepal, Kathmandu, Nepal, 2011.

[49] UNFCC. Second National Communication Report of Nepal to UNFCCC. Ministry of Science Technology and Environment, Kathmandu, Nepal, 2015.

[50] IPCC. Climate Change 2007: Synthesis Report. Contribution of Working Groups I, II and III to the Fourth Assessment Report of the Intergovernmental Panel on Climate

Change. Core Writing Team, Pachauri, R.K. and Reisinger, A. (eds.). IPCC, Geneva, Switzerland, 2007, 104 pp.

[51] Bajracharya, S.R., S.B. Maharjan, and B.R. Shrestha. Second Generation Glaciers Mapping and Inventory of Nepal. Journal of Nepal Geological Society (Sp. issue-Abstract), 2010, 41 (21).

[52] ICOLD. ICOLD Bulletin on Dam Safety Management. s.l.: International Commission of Large Dams, 2007.

[53] MoHA. Nepal Disaster Report 2015. The Government of Nepal, Ministry of Home Affairs (MoHA) and Disaster Preparedness Network-Nepal (DPNet-Nepal), 2015, 127p.s

[54] J-Rapid Final Workshop 21 June, 2016."Impact on Infrastructure by Gorkha Earthquake 2015 Induced Landslides" Inventory Mapping of Landslides Induced by the Gorkha Earthquake 2015 and a Proposal for Hazard Mapping of Future Landslides for Making. Kathmandu, s.n., 2016.

[55] GoN. Hydropower Development Policy 2001 (2058 BS). GoN, Ministry of Water Resources, Kathmandu, 2001.

[56] Rosenberg, D.M., V.H. Resh, S.S. Balling, M.A. Barnby, J.N. Collins, D.V. Durbin, T.S. Flynn, D.D. Hart, G. A. Lamberti, E.P. McElravy, J. R. Wood, T.E. Blank, D.M. Schultz, D.L. Marrin, and D.G. Price. Recent Trend in Environmental Impact Assessment. Canadian Journal of Fisheries and Aquatic Science : Journal of Fisheries and Aquatic Science, 1981, 38, 591–624.

[57] Beanlands G.E. and Duinker P.N. An ecological framework for environmental impact assessment. Journal of Environmental Management, 18: 267–277, 1984.

[58] Beanlands, G.E. In Situ Contaminants and Environmental Assessment-A Ecological Summary. s.l.: Hydrobiologia, 1987, 149, 113–118.

[59] Schaeffer, D.J., E.E. Herricks, and H.W. Kerster. Ecosystem Health: I. Measuring Ecosystem Health. s.l.: Environmental Management, 1988, 12, 445–455.

[60] WECS. National Water Plan 2005, Nepal. Water and Energy Commission Secretariat, Kathmandu, Nepal, 2005.

[61] NEA. Annual Report. Nepal Electricity Authority, Kathmandu, Nepal, 2008, 108 p.

[62] Chaulagain, N.P., Impacts of Climate Change on Water Resources of Nepal: The Physical and Socioeconomic Dimensions. s.l.: Shaker Verlag, Aachen, Germany, 2007, 146 pp.

[63] Chaulagain, N.P. Socio-economic Dimension of Snow and Glacier Melt in the Nepal Himalayas. Dynamics of Climate Change and Water Resources of Northwestern

Himalaya. s.l.: Society of Earth Scientists Series, 2015. DOI 10.1007/978-3-319-137 43-8_15.

[64] IHA. Hydropower Status Report 2016. International Hydropower Association, Sutton, London, UK, 2016, 41 p.

[65] NEA. Annual Report: A-Year-in-Review-FY-2012-13. Nepal Electricity Authority, Kathmandu, Nepal, 2013.

[66] DoED. Guidelines for Study of Hydropower. Department of Electricity Development (DoED), Kathmandu, Nepal, 2003.

[67] Reynolds, J. Assessing Glacial Hazards for Hydro Development in the Himalayas, Hindu Kushand Karakoram. s.l.: Hydropower and Dams, 2014, 2, 60–65.

[68] NPC. Sustainable Development Goals, 2016-2030. s.l.: Government of Nepal, National Planning Commission, Kathmandu, Nepal, 2015.

[69] WEC. World energy resources. World Energy Council. charting the upsurge in hydropower development 2015. WEC, London, UK, 2015.

[70] Kumar, Bajracharya Samjwal Ratna and Mool Pradeep. Impact of global Climate Change from 1970s to 2000s on the Glaciers and Glacial Lakes in Tamor Basin, Eastern Nepal. ICIMOD, Kathmandu, Nepal, 2006.